THE VEGAN EVOLUTION

Arguing for a vegan economy, this book explains how we can and should alter our eating habits away from meat and dairy through sociocultural evolution.

Using the latest research and ideas about the cultural ecology of food, this book makes the case that through biological and, especially, cultural evolution, the human diet can gravitate away from farmed meat and dairy products. The thrust of the writing demonstrates that because humans are a cultural species, and since we are evolving more culturally than biologically, it stands to reason for health and environmental reasons that we develop a vegan economy. The book shows that for many good reasons we don't need a diet of meat and dairy and a call is made to legislative leaders, policy makers, and educators to shift away from animal farming and inform people about the advantages of a vegan culture. The bottom line is that we have to start thinking collectively about smarter ways of growing and processing plant foods, not farming animals as food, to generate good consequences for health, the environment, and, therefore, animals. This is an attainable and worthy goal given the mental and physical plasticity of humans through cooperative cultural evolution.

This book is essential reading for all interested in veganism, whether for ethical, environmental, or health reasons, and those studying the human diet from a range of disciplines, including cultural evolution, food ecology, animal ethics, food and nutrition, and evolutionary studies.

Gregory F. Tague is a Professor in the Departments of Literature, Writing and Publishing and Interdisciplinary Studies and founder and senior developer of The Evolutionary Studies Collaborative at St. Francis College, New York, USA. He is also the founder and organizer of a number of Darwin-inspired Moral Sense Colloquia and has written numerous books, including most recently *An Ape Ethic and the Question of Personhood* (2020), *Art and Adaptability: Consciousness and Cognitive Culture* (2018), *Evolution and Human Culture* (2016), and *Making Mind: Moral Sense and Consciousness* (2014).

ROUTLEDGE STUDIES IN FOOD, SOCIETY AND THE ENVIRONMENT

Resourcing an Agroecological Urbanism
Political, Transformational and Territorial Dimensions
Edited by Chiara Tornaghi and Michiel Dehaene

Food, Senses and the City
Edited by Ferne Edwards, Roos Gerritsen and Grit Wesser

True Cost Accounting for Food
Balancing the Scale
Edited by Barbara Gemmill-Herren, Lauren E. Baker and Paula A. Daniels

Holiday Hunger in the UK
Local Responses to Childhood Food Insecurity
Michael A. Long, Margaret Anne Defeyter and Paul B. Stretesky

School Farms
Feeding and Educating Children
Edited by Alshimaa Aboelmakarem Farag, Samaa Badawi, Gurpinder Lalli and Maya Kamareddine

The Vegan Evolution
Transforming Diets and Agriculture
Gregory F. Tague

For more information about this series, please visit: www.routledge.com/books/series/RSFSE

THE VEGAN EVOLUTION

Transforming Diets and Agriculture

Gregory F. Tague

LONDON AND NEW YORK

Cover image: Getty Images

First published 2022
by Routledge
4 Park Square, Milton Park, Abingdon, Oxon OX14 4RN

and by Routledge
605 Third Avenue, New York, NY 10158

Routledge is an imprint of the Taylor & Francis Group, an informa business

British Library Cataloguing-in-Publication Data
A catalogue record for this book is available from the British Library

Library of Congress Cataloging-in-Publication Data
A catalog record has been requested for this book

ISBN: 978-1-032-26764-7 (hbk)
ISBN: 978-1-032-26762-3 (pbk)
ISBN: 978-1-003-28981-4 (ebk)

DOI: 10.4324/9781003289814

Typeset in Bembo
by Taylor & Francis Books

As always, for Fredericka and Karolina

CONTENTS

ACKNOWLEDGMENTS

While the initial draft of this book was written early in the pandemic year 2020, it is not a work produced in isolation. Many discussions with Fredericka A. Jacks moved the project along. I was inspired by the written words of numerous people, reflected in the bibliography, but noteworthy is Carlo Alvaro. The Heterodox Academy provided a forum for comments and criticism, sponsoring my talk on veganism, 9 July 2020. My inclusion in the Eighth International Conference on Sustainable Development panel, 21 September 2020, chaired by Dr. Meredith Storey, helped shape some ideas here. My work on the Ninth International Conference on Sustainable Development, September 2021, with my co-author Sintia Molina, also proved valuable. I'd like to express my deep appreciation for a number of ape sanctuary personnel who responded to a diet survey. I benefited from webinars like "Animals as Drivers of Climate Change," 9 December 2020 and "The Future of Animal Politics," 21 January 2021, organized by Eva Meijer, which confirmed and refined some of my thinking. A webinar of 5 May 2021 on animal advocacy organized by Lori Marino of the Kimmela Center featuring Barbara J. King and Amanda Stronza proved helpful. Attending meetings of the LEAP Group on Sustainability organized by Isabel Rimanoczy was a great benefit. Hearing others or engaging with them via email and across social media added to my thoughts, and I'd have to mention, naming only a few, people like Jeff Sebo, Josh Milburn, Jennifer Verdolin, Natalie Khazaal, Marc Bekoff, David Steele, Elisa Aaltola, Lesley Newson, and Gary L. Shapiro. Academic works are creative enterprises and draw from a number of outside sources. The voices I'm fortunate to work with in editing *Literary Veganism: An Online Journal* have had a meaningful effect on my ideas and feelings. I'm also indebted to anonymous readers for valuable feedback. I'd especially like to thank my editor, Hannah Ferguson, at Routledge, for steering this book to publication.

SIMPLIFIED CHART OF PRIMATES AND DATES

Sedentism and agriculture, 11–8kya
Upper Paleolithic/Late Stone Age 36kya–10kya
Anatomically Modern Humans in Europe, 40kya
Global diaspora of Anatomically Modern Humans, 70kya
Earliest Anatomically Modern Human fossils, Africa, 200–300kya
Middle Paleolithic/Middle Stone Age/Mousterain, 300–30kya
Global diaspora, *Homo erectus*, 1.7mya
Modern cranial capacity, 250kya
Homo neanderthalensis, 250–28kya
Homo neladi, 335kya
Archaic *Homo sapiens* and *Homo heidelbergensis*, 500–100kya
Homo erectus, 1.8mya–300kya
Homo habilis, H. rudolfensis, and *H. ergaster*, 2.5–1.6mya
Oldowan stone tools, 2.6–1.8mya
Gracile and Robust Australopiths, 4.4–1.2mya
Ardipithecus ramidus, 4.5mya
Early *Homo* split from predecessor common to Chimpanzee 6–7mya
Chimpanzee split from predecessor 6–7mya
Bonobo split from Chimpanzee 2.5mya
Gorilla split from predecessor 7–9mya
Orangutan split from predecessor 16–20mya
Baboon, Macaque, Vervet with origins about 26–36mya
Lemur and Tarsier types, before 45mya
Pleistocene, 1.8mya
Pliocene, 5mya
Miocene, 23mya
Oligocene, 38mya

Eocene, 54mya
Paleocene, 65mya
Cretaceous Period, 145mya
[Code: mya = million years ago; kya = thousand years ago]

OVERVIEW

Veganism in some forms can be traced back to Biblical times (e.g., the book of Daniel in the Old Testament) and is alive in, for instance, an ancient Indian religion like Jainism. Vegans avoid harming or eating animals or exploiting animals for their skin, fur, or bodies. Other than for religious reasons, vegans are motivated by diet or health, ethical concerns, psychosocial responses to celebrities or fads, or identity politics in the form of activism. Most corporations and many people see no profit in ethics, so the weight in this argument for ethical veganism falls on establishing the resilience and sustainability of human and environmental health. Issues include uneven food consumption, collective implications of animal farming, and personal gain over community ecology. Corporations and entrepreneurs are capitalizing from a plant-based trend, but often their actions are not fostering the conservation but the exploitation of resources. Minds, eyes, voices, and hands should be on how a vegan economy across industrial nations can prevent poor health and mitigate climate change. Certainly, ethics are constituents of sustainability goals, as the United Nations is well aware. For developing countries where food instability is a concern, wealthy nations could help them adapt to veganism in the wake of global warming without relinquishing cultural beliefs or practices. Technology and laws are not primary solutions for achieving a healthy environment. Recycling is not of itself a final solution. Energy loss and food waste, especially from animal agriculture, must be eradicated. This is an argument demonstrating how in our ancestral hominin lineage we were plant and fruit eaters, just like our living relatives, the great apes. Biologically, we can survive on a plant-based diet. More so, with the mechanism of cultural evolution, the arts as much as the natural and social sciences can educate young people about the benefits of a vegan culture to generate advantageous shifts in attitudes about physical, environmental, and animal well-being.

PREFACE

This book is about the human diet: what it was, how it changed, and its power to transform health, norms, and the environment for years to come. In a few words, the book covers the ecology of food culture. In biology, organisms have relations to each other in their surroundings. This is true, too, for humans who manipulate, sometimes grossly, ecosystems to farm meat and dairy products. I hope to show how our current food ecology is detrimental to our physical health, the earth's climate, and animal lives. Drawing from primatology, paleoanthropology, biology, cultural evolution, and other disciplines the purpose is to demonstrate how physical and global wellness can be achieved through education and awareness efforts geared toward a vegan culture. Relying less on technology and more on informative shifts in attitudes, values, and beliefs we can effect positive alterations in our ecology of food. Organizations like the Factory Farming Awareness Coalition (2021) are driving such change. Advances in climate and agricultural technology should follow, not lead, an ecologically and animal friendly frame of mind.

On the most basic level, here are some subjects tackled and for which some theorizing will be put forward throughout the book.

1. We do not need to consume animal products like meat, seafood, and dairy.
2. If we stopped eating meats and dairy, we'd still survive physically and economically.
3. While infrastructure and social changes are required in our food ecology, individuals and groups need to fit into a new mindset.
4. Veganism is a sustainable solution, not a passing digression, and it's feasible in educational institutions, business and government settings, and local communities.
5. Cultural evolution is a type of evolutionary mechanism to engender a healthful and ethical transformation in the social ecology of food.

This is an argument that includes analogies. This is not a policy handbook. While I draw on the facts of biology and human evolution from standard and current sources, my reference to biology (Chapter 2), apes (Chapter 3), and australopiths (Chapter 4) is mainly to loop together various disciplines into cultural evolution. This is not strictly a text about selection in human evolution; rather, sources from science disciplines are used to support the claims in the argument. While biological evolution is important, the thrust of this book is about cultural evolution (Chapter 5). Without making a generalization, some aspects of being human are more cultural in an ongoing present than natural in a genetic past. There is, nonetheless, continuity and overlap as will be shown. Not all traits reliably reproduce or get copied, and some are selected out. Genes and culture are not so easily separable since selection can operate on a number of biological and cultural levels. We should have learned by now that immediate material benefit is not a strategy for any long-term solution.

The main audience would include, but is not limited to, well-fed urban and suburban people among wealthy, industrialized nations since they have the funds and tools for change, as opposed to poor people in developing nations whose food security is at risk. I'm especially keen to reach young people because their future is at stake. Wangari Maathai (2006), Nobel Peace Prize recipient and creator of the Green Belt Movement, talks about how school gardens in her country of Kenya helped wean students from unhealthy fast foods. More so, industrialized countries have an obligation to help developing nations secure a stable plant food system that does not destroy local farms or ecosystems. Maathai points out how colonizers disabled the self-reliance of indigenous people in relation to their own, precious environment by making them see land and resources as commodities for exploitation. Likewise, Vandana Shiva, an Indian ecofeminist fights for the rights of communities to have control over natural resources, not multinational corporations that enter India (or Africa or Indonesia) to plunder the riches of biodiversity, whether animals, minerals, land, or woodlands, for profit. Thinkers like Maathai and Shiva make clear that we need to change course. Karen Warren (2000, 2015) notes that in Africa or India subsistence and eco-friendly farming by women is often replaced by unsustainable monocultures initiated and managed by corporate men. Even urban areas, which abuse natural assets near and far, can adopt a vegan culture rooted in a network of communities as a certain beginning. Pushing this thought further, there could be bias since vegan or animal studies can be, according to some scholars, mediated by privileged, colonial attitudes of white, Western culture (Deckha 2018).

Returning to the central tenets of this book, ignorance about our evolutionary past is filled with gaps rendering the completion of different scenarios inadequate. I acknowledge our lack of a complete picture regarding our hominin past, but what's more crucial is using data we have to build a better future. In the wild, animal lives matter because they connect to sustainable ecosystems. Farming livestock for food has proved otherwise and is responsible for over 50 percent of harmful emissions (Goodland & Anhang 2009). Food replacements not originating from meat or dairy farming would have a large and immediate effect. Still, many conservative ideologies express a heightened bias against vegans, perceiving them as a threat and believing vegan motivation is for animal or environmental causes and not health (MacInnis & Hodson 2017). Research shows how people who identify as conservative tend to adhere to their beliefs in spite of evidence to the contrary (Pennycook, et al. 2019). Crucial issues of good health and an uncontaminated environment should always transcend religious or political ideologies. The hope is that in reading this book, answers, as well as valuable questions, will be presented and that any prejudices will be mitigated so that some people can cross over to join a vegan culture. The invitation to be vegan is always open.

In some respects, with its turn away from privileging "humanity," *The Vegan Evolution* could be read as a footnote to my book *An Ape Ethic and the Question of Personhood* (2020), since here, more directly for humans, I'm favoring collective action for environmental health and sustainable development. There were several factors for writing this book. In late December, 2019, Fredericka Jacks and I had started *Literary Veganism: An Online Journal*. Then, SARS-CoV-2 struck (hereafter the pandemic, the coronavirus, or simply the virus). As ethical vegans we were quite upset that, outside of some chatter by vegan groups and animal studies academics, there was little serious discussion in the mainstream media about the cause of the virus and how to mitigate another outbreak. The origin of the pandemic in human slaughter and eating of endangered or rare species was known among scientists for years (Pew Commission on Industrial Farm Animal Production 2008). After the fact, politicians in the United States and leaders in China worked on banning wet markets. However, an article in a major newspaper (Severson 2020) predicts how we would eat differently in a post-pandemic world, and not unexpectedly it's still mostly meat and dairy.

Animals do not need to be human food. Human consumption of animal products, whether flesh or dairy, equates to unnecessary cruelty and death. That's not a typical or popular position. The focus, however, will not be on animal rights specifically since that was covered to some extent in *An Ape Ethic*. Nevertheless, it's difficult to write about veganism and not conjure the long shadow of animal persecution. Jim Mason (2005) suggests that humans need to recognize how central animals are to the natural world and not objects for us to destroy, either in their habitats or factory farms. Animals are not fictional objects, but subjects of a life (Regan 2004), agents who value being alive, clear in how some escape the slaughterhouse in an instinct for self-preservation (Colling 2020). Cows are, much like sheep, pigs, etc. peaceful, so imagine the terror of any creature who sees and hears the wails of death coming from others.

It's difficult to say this is not an ethical argument, but I don't want to dwell on moral abstractions or theoretical examples. A good illustration of that would be John Hill's *The Case for Vegetarianism* (1996). Many of his points are still, regrettably, valid. Some readers will present counter claims that any argument for veganism is too difficult to achieve, personally or economically. Hill addresses similar objections. First, it is easy to transition from meat and dairy to whole foods and vegan products, if done properly and gradually. Second, there's a cultural opportunity in veganism to invest in local, urban vegetable farming and the manufacture of minimally processed vegan foods. Like other authors, Hill covers the bad consequences of livestock production, needless animal cruelty, poor health, pollution, and world hunger. Most of that is beyond the scope of this book, and though at times understated, there is moral significance to the argument for veganism.

Instead, by looking at the past and present perspectives, through biological and cultural evolution, goals for the near future will become focused. We need to concentrate on the prevention of disease and the cessation of global warming and not simply on diagnoses, pharmaceutical treatments, or short-term corrections. Unavoidably, there are moral questions and ethical dilemmas (one's feelings about moral norms) in these concerns. For instance, an examination of *The Moral Complexities of Eating Meat* (2016), edited by Ben Bramble and Bob Fischer, is coming up. The meat you eat from your plate was an individual personality with a mind and emotions, as Barbara King (2017) eloquently argues. Animals are raised cruelly for slaughter to be mutilated and ingested by humans, causing bad health, disease, as well as environmental problems. It becomes morally complex when we start spinning abstractions to justify our actions. Hill (1996), too, dwells on religious and metaphysical questions. My area is more with the evolutionary components of meat and dairy eating, not the "philosophical." Certainly, economic, social, and moral ideas are either stated or implied in my argument, one grounded in the basics of science. This is not to minimize the importance of philosophy, which can indeed alter the course of human development. Consider, for instance, in line with the thinking here, how some people in political power and economic control ignore wise counsel and overlook injustice if they financially benefit.

There have been a few other books tangentially related to the subject under consideration. Josh Berson's *The Meat Question: Animals, Humans, and the Deep History of Food* (2019) claims that meat does not have to be in any human diet. Nevertheless, his book seems more about the demand for meat and what drives that need. On the question of whether or not meat was essential for early *Homo*, Berson says that by the time of *Homo erectus* the human diet improved in its versatility, but any absence of meat at that stage did not impair physiology or neurobiology. That outlook seems to privilege the carnivory of *Homo sapiens* over other early species and living apes. Berson's book is not an evolutionary argument for veganism, nor is Benjamin Wurgaft's *Meat Planet: Artificial Flesh and the Future of Food* (2019). In fact, Wurgaft does not seem to make claims in an argument. His book is observation, investigation, questions, and reflections about in-vitro lab meat. The so-called meat question, which both books push into the future, poses a paradox. The animal farmers, purveyors, and eaters of meat are the ones who claim, or we should say promote, the expanding need and demand for meat. This is an illusion. There really are no firm physical, psychological, economic, or cultural barriers preventing us from shifting to a vegan economy.

Some demands on the ethics of readers are made since personal action is advocated. Morality and ethics are not solely about what is good or right for humans alone. We have moral and ethical obligations to ecosystems and all animals wild or domestic. Human freedoms and liberties do not extend to wanton disregard of other life forms, as virtue ethicist Rosalind Hursthouse (2000) notes. The ethical citizen considers the best outcomes regarding health and the environment, not necessarily in large numbers, but for human and animal individuals. In this way, it's difficult to pinpoint any moral theory used here. Instead, I look to moral individualism that could bring about good consequences for the entire biosphere. Philosopher James Rachels (1990), against utilitarian principles, calls for a focus on individual identity and autonomy. Willful actions, outside of mere self-interest, and not just consequences are important. Moral individualism affirms the ethical state of the individual in relation to others in a shared ecology. A utilitarian or a deontologist might focus on acts and not the actor. A utilitarian (Singer 2009) sees the moral measure of an act in its larger consequences. For a deontologist (Regan 2004), some acts are right, some wrong, no matter the consequences. Bearing witness to the atrocities inflicted on animals by humans is consequentialist, in terms of learning how not to do this again, and moral, regarding the rights of an individual life (Jenni 2018). Mine is not a morally or culturally relativist argument. Motives and results matter. Consumption of meat and dairy produced on any scale contributes to poor health, causes environmental pollution, and persecutes animals (Stone 2011). This is fact-based information dripping with evaluative observations, which some people choose to ignore. Maybe the evolutionary argument of this book can change a few minds.

Philosopher Lucy Schultz (2020) points out how, traditionally, nature has been set off in contrast to culture. There was a time when we were part of nature. Then, with the formation of cities, human culture separated from and eventually, with industrialization, invaded nature. One wonders if nature, as perceived by the British Romantic poets epitomized in William Wordsworth, even exists any longer. Barbara King (2021) suggests that people, who have disconnected from nature via modern, industrial life and technology, could learn something about themselves if more time were spent reflecting in nature. This practice in children could cultivate feelings of compassion for other species, whether mammals or insects, observing their evolved behaviors and culture. The trick is not simply to observe, but to experience the lives of other creatures. Children are "most susceptible" to marketing of food products, so it's a global concern to shift their attention away from meat and dairy to plant-based products (Goodland & Anhang 2009). At the same time, Schultz goes on, humans are biological animals derived from and evolved through nature. We've fabricated possessions from nature, but not to every living creature's advantage. Western art and philosophy created a false idea of "nature" as wildlife, a landscape, another world, etc. This flawed dichotomy, like the erroneous separation of mind and body promulgated by Descartes, has enabled humans to justify their abuse of animals and ecosystems.

In reality, now evident from the forceful onset of climate change, humans are dependent on and subject to, by their own excesses, the energies of nature. We need to stop competing with the balances in the natural world and cooperate, even on the bodily level. If not misinterpreting him, biologist Richard Alexander (2005) says that culture emerges from ontogeny. Culture is not unconnected to nature. Our primate genealogy is one of ancient kinship with the natural world and its ecology of food. The human relationship with nature can and must be restored through a culture of veganism.

References

Alexander, Richard D. 2005. "Evolutionary Selection and the Nature of Humanity." *Darwinism and Philosophy*. Vittorio Hösle and Christian Illies, eds. Notre Dame, Indiana: U Notre Dame P. 301–348.

Berson, Josh. 2019. *The Meat Question: Animals, Humans, and the Deep History of Food*. Cambridge, MA: MIT P.

Bramble, Ben and Bob Fischer, eds. 2016. *The Moral Complexities of Eating Meat*. Oxford: Oxford UP.

Colling, Sarat. 2020. *Animal Resistance in the Global Capitalist Era*. East Lansing, MI: Michigan State UP.

Deckha, Maneesha. 2018. "Postcolonial." *Critical Terms for Animal Studies*. Lori Gruen, ed. Chicago: U Chicago P. 280–293.

Factory Farming Awareness Coalition. 2021. https://ffacoalition.org/.

Goodland, Robert and Jeff Anhang. 2009. "Livestock and Climate Change." *World Watch*. https://awellfedworld.org/wp-content/uploads/Livestock-Climate-Change-Anhang-Goodland.pdf.

Hill, John Lawrence. 1996. *The Case for Vegetarianism: Philosophy for a Small Planet*. Lanham, MD: Rowman and Littlefield.

Hursthouse, Rosalind. 2000. *Ethics, Humans and Other Animals*. London: Routledge.

Jenni, Kathie. 2018. "Bearing Witness for the Animal Dead." *Proceedings of the XXIII World Congress of Philosophy* 12: 167–181.

King, Barbara J. 2017. *Personalities on the Plate: The Lives and Minds of Animals We Eat*. Chicago: U Chicago P.

King, Barbara J. 2021. *Animals' Best Friends: Putting Compassion to Work for Animals in Captivity and in the Wild*. Chicago: U Chicago P.

Maathai, Wangari. 2006. *The Green Belt Movement*. New Revised Edition. New York: Lantern Books.

MacInnis, Cara C. and Gordon Hodson. 2017. "It Ain't Easy Eating Greens: Evidence of Bias Toward Vegetarians and Vegans From Both Source and Target." *Group Processes and Intergroup Relations* 20 (6): 721–744. https://doi.org/10.1177/1368430215618253.

Mason, Jim. 2005. *An Unnatural Order: Why We are Destroying the Planet and Each Other*. New York: Lantern Books.

Pennycook, Gordon, et al. 2019. "On the Belief That Beliefs Should Change According to Evidence." *PsyArXiv*. https://doi.org/10.31234/osf.io/a7k96.

Pew Commission on Industrial Farm Animal Production. 2008. "*Putting Meat on the Table: Industrial Farm Animal Production in America*." https://www.pewtrusts.org/en/research-and-analysis/reports/0001/01/01/putting-meat-on-the-table.

Rachels, James. 1990. *Created from Animals: The Moral Implications of Darwinism*. Oxford: OUP.

Regan, Tom. 2004. *The Case for Animal Rights*. Updated edition. Berkeley, CA: U California P.

Schultz, Lucy. 2020. "Climate Change and the Historicity of Nature in Hegel, Nishida, and Watsuji." *Environmental Philosophy* 17 (2): 271–290. doi:10.5840/environphil2021127101.

Severson, Kim. 2020. "How Will We Eat in 2021? 11 Predictions to Chew On." *The New York Times* 22 December.

Singer, Peter. 2009. *Animal Liberation*. Updated 1975 edition. NY: Ecco.

Stone, Gene, ed. 2011. *Forks Over Knives: The Plant-based Way to Health*. NY: The Experiment.

Tague, Gregory F. ed. *Literary Veganism: An Online Journal*. www.litvegan.net.

Tague, Gregory F. 2020. *An Ape Ethic and the Question of Personhood*. Lanham, MD: Lexington Books.

Warren, Karen. 2000. *Ecofeminist Philosophy*. Lanham, MD: Rowman and Littlefield.

Warren, Karen. 2015. "Feminist Environmental Philosophy." *Stanford Encyclopedia of Philosophy*. Plato.stanford.edu

Wurgaft, Benjamin Aldes. 2019. *Meat Planet: Artificial Flesh and the Future of Food*. Oakland, CA: U California P.

INTRODUCTION

Eating Animals is Bad for Health and the Environment

Building from the Preface, let's outline some areas this argument will more comprehensively cover in upcoming chapters, not necessarily in this order but over the course of the discussion in a dialogic fashion. These broad areas are interrelated in the overarching argument and claims of this book. There are many moving parts here, but let's rephrase the controlling thesis. Because we evolved as plant-based eaters, still evident in our cousins the great apes, and since that diet is healthy for individuals and ecosystems, we should employ education and cultural evolution to raise awareness about the profits of veganism.

Paleoanthropology

The diversity of human dietary history tells us that it's biologically feasible to establish a vegan culture. As omnivores related to mostly herbivorous great apes, we don't need to eat animals. We could at least cut back such consumption dramatically. Highly processed animal products (containing, e.g., sodium nitrate) cause health problems. Generally speaking, most animal agriculture is on a large scale and is environmentally unsustainable. There are ethical concerns, too, in the farming and eating of animals.

Culture

Understanding how and why human culture evolves informs the goal of promoting a vegan economy. Our species is evolving more in terms of culture than biology. Apes have culture, and we share a common ancestor with them. Culture is ideas, values, beliefs, practices, and technology. We rely on culture to survive and it shifts regularly. Some cultural tendencies can be detrimental, such as the oppression of people or immoderate consumption of unhealthy foods.

DOI: 10.4324/9781003289814-1

Education

Young people will be the linchpin to a cultural shift that values good health, a stable environment, and a world free of cruelty to animals. Learning is not indoctrination, rather it is part of the mechanics of cultural evolution. Informing young people, who will inherit and govern the future, about mistakes made by previous generations will be a critical step in cultural development to personal and environmental health. The literary, visual, and musical arts could potentially be transformational in this effort.

Communities

As part of this cultural shift, individuals could become ethical vegans by choosing virtues of temperance and moderation in food and material consumption and by practicing kindness toward animals. Communities working with ethical vegans can regulate existing infrastructure to educate young people about the benefits of a vegan culture. Local governments can utilize existing structures to establish community green gardens and kitchens as workshops to learn about and how to cook vegan food. Success in adopting a vegan culture on the community level could then be brought up to the state and federal levels in any country.

Ethics

Humans tend to be a caring and helping species. We value health and family. We hold dearly to moral traditions. Every culture has values and beliefs. In the vision of a vegan culture, the essential human *mores* of community support and the ethics of avoiding harm to others could be filtered into a vegan outlook that values plant food over meat and dairy. In turn, an ethos of care for the environment and animals will flourish. Alice Crary (2018) offers an unsentimental overview of ethical approaches to life by asking if values exist in the natural world. As with humans, animal lives are not intellectually or morally neutral, revealed through observation of behaviors.

Economy

A vegan culture could grow a vegan economy, and this is desirable since the advantages in time outweigh the costs. Gains will be made in health to many people and the environment. Considering the fitness loss and environmental damages from industrialization's food contaminations and ecological pollutions, a reasonable goal is to have our children and their offspring adopt a healthier lifestyle that can break cycles of disease, climate change, and animal cruelty.

Inclusivity

While initially there will be significant resistance to veganism, efforts in education and community awareness will bring acceptance. In fact, that is exactly how culture changes and evolves. To what extent will people in developed nations be willing to alter their diets, and how quickly, are determining factors blending into a vegan economy. Many people can gradually become vegan over the course of about one year. Whoever accepts a plant-based and animal-friendly diet will realize that vegans are not strange beings from another planet, but are compassionate, eco-friendly individuals in a socially responsible community.

Ecofeminism

Feminist environmental philosophy does not inform this book but is at least on the margins since issues of domination of the environment often include women, especially those of color and in lower economic classes. Deforestation to raise livestock or establish a palm oil plantation in addition to water pollution from factory farming are environmental problems that impact all people, but especially women who head a household and usually in communities of color. Sexist men tend to see women, the environment, and animals as inferior or property to be exploited (Gruen & Weil 2012). Ecofeminists attempt to break the dualism of oppressive constructs, like male/female or human/nonhuman, and focus on care (Gruen 2011).

The design of this narrative is straightforward and fits a standard pattern. The opening pages through the first chapter present a number of pertinent ideas and issues. This relevant miscellany is subsequently charted by chapters that get into the details of biological evolution, great ape diets, what and how our ancestral relative australopithecines ate. Then, there's a keystone chapter on the mechanism of cultural evolution followed by a capstone conclusion. In a daring way, this is a work of synthesis, where multiple disciplines cross, integrate, and work together under the broad umbrella of cultural evolution.

Layout of Topics Under Consideration

Here's a sketch of the argument and claims that will proceed to unravel.

1. We are evolved animals.
2. We evolved in an environment that favored our eating plant foods.
3. We opportunistically scavenged meat from carcasses during food scarcity in prehistory.
4. We share morphology, physiology, and behaviors across extant great ape species.
5. We don't need to eat copious amounts of meat as evidenced in great apes.
6. We evolved culture in ways unlike great apes, for example in farming.
7. We employed cultural evolution over genes to enhance our survival and reproduction.
8. We started to become sedentary around 10kya and formed cities around 8kya.
9. We controlled and stored food at that time through agriculture.

10. We conquered the earth as our populations and cultures exploded and varied.
11. We fashioned massive farms to harvest meat and dairy for denser populations.
12. We degraded local and distant environments with industrialized farming.
13. We promoted a corporate culture of meat, seafood, and dairy eating.
14. We need to consider a shift to veganism for human and biotic health.
15. We can service cultural evolution advantageously with awareness to change eating habits.
16. We can solicit visionary leaders, educators, and artists to promote veganism.

Ernst Haeckel (1879), in summarizing Charles Darwin's (1859) idea of descent by modification, says that in the struggle for existence new species evolve without design; human willpower, on the other hand, cultivates variety with design. Genetic evolution has enabled cultural capacities cognitively; in turn, the brain psychologically adapts to a changed cultural milieu. This process alters existing cultural norms or artifacts while the brain's neural connections ratchet up cultural innovation with feedback loops. Some cultural forms are destructive, some beneficial. Meat and dairy production and consumption are malefic practices we can evolve away from culturally.

A large part of this project involves education. With knowledge that is properly informed and balanced by science, people are able to make moral decisions and change their habits. Of course, one has to be educable and willing to alter his habits. I cannot, therefore, reach the minds or change the attitudes of everyone, but the anticipation is that some will follow the lead offered in this text and consider a vegan way of life. While this work is theoretical, there are practical applications. I do not offer a blueprint of how a vegan economy would work, but some outlines have already materialized. For a pragmatic strategy see Tobias Leenaert's (2017) *How to Create a Vegan World*. Like Barbara King (2021), Leenaert advocates a relaxed version of veganism or reducetarianism where people still consume animal flesh, cheese, butter, milk, etc. but on a reduced scale. A physician would not advise someone to smoke only half a pack of cigarettes a day. Considering the billions of hungry consumers, reduction of meat and dairy intake is a weak and not an optimal solution. Paradoxically, because of the modern human demand for so many animal products the climate could become less stable for vegetable farming. Michael Clark, et al. (2020) say that reliance on animal farming, coupled with poor land management and wastefulness, will prevent humans from reaching climate change targets. Marco Springmann, et al. (2018) earlier predicted this dire evaluation.

Consider that irony. Even with a shift to vegan agriculture, if we don't move fast enough global warming from producing so much livestock for food (IPCC 2021) might hinder our attempts to farm plants successfully in some places. Nurturing chickpeas, soybeans, and lupines is more agriculturally sustainable than grains for cattle (Di Paola, et al. 2017). This shift won't be easy in spite of the urgency. Unfortunately, the Pew Research Center (IPCC 2018) says that U.S. Americans are more concerned about cyber or terrorist attacks than climate change, though that has likely changed in the past few years. At the same time, Pew (Tyson & Kennedy 2020) reports that most Americans across political lines think the government should do more regarding the climate, like tree planting, carbon capture tax credits, and carbon taxes on corporations. While any state or federal government, in whichever country, can and should promote the security and health of its citizens, on an individual and community level there are plenty of actions one can take, as suggested here, such as embracing and advertising a vegan culture.

On a local level we tend to cooperate with those we know. The biological precedents for the puzzle of cooperation are in reciprocity and kin relationships. However, I'd counter by noting that we incline to be a self-interested species. As one example, during the 2020 pandemic many people in some developed countries simply refused to wear a face mask knowing it would help prevent the spread of the virus. This does not mean we don't care about or help others, as was advantageous in our prehistory. Not to sound cynical, but because of better health care and longer life expectancy in many industrialized nations some people and groups who are selfish survive and pass on their genes and values. What's happened over time with explosions in population and industry is that competition seems to reign.

Capitalism is at the root of today's speciesism and anti-climate society. The fascination with competition is not just from evolutionary theory but a result of capitalist culture. During the height of the 2020 pandemic, while there were heroic acts on many individual and small group levels, the federal government in the United States was slow to respond, which set states in competition with each other for basic protective gear, medicines, and machinery. Although this is simplified, eventually the government almost ignored cooperating with the states. Businesses looking for fat contracts competed with each other to satisfy the sudden need for medical supplies. E-commerce corporations raked in profits. Vendors and delivery companies competed with each other as families forced into lockdown required and bid for food and other necessities. Some people greedily stockpiled goods.

This is why we won't see any sudden, national shift to a vegan economy in spite of the urgency. It must start on a community level and spread from there as a beacon of cooperation, seen in examples of neighborhood vegan food cooperatives and schoolyard gardens. Of course, without policy change it's unlikely there will be systematic shifts, and so there's a debate about top-down or bottom-up social change. In what I'm suggesting, there could be a combination of both. Under normal circumstances, restaurants compete with each other as they strive to fill their tables with hungry omnivores. Across the globe there are any number of cattle farmers and meat purveyors competing with each other to satisfy the growing demand for animal flesh as already well-fed populations in concentrated areas increase. Corporations vie for consumers to snatch up products. Politicians lie to out-compete one another. Nations fight for sea and land. Individuals, groups, and countries kill those they don't want or don't like. There's more mutual interdependent activity among nonhuman species. At the same time, some research (Handley & Matthew 2020) suggests that competition between groups can fuel cooperation through cultural group selection. Joseph Henrich (2016), too, says that competition between groups drives cultural evolution.

Certainly, harmony can evolve in groups and on the edges between groups, with new cultural combinations and copying, which is the point I'm striving to make. Ecofeminists, deforestation fighters, ethical vegans, Black vegans, raw vegans, reducetarian dieters, climate activists, factory farm reformers, animal ethics advocates, nutrition educators, animal or environmentally friendly artists (whether in writing, music, or the visual arts), animal rights lawyers, animal or vegan studies scholars, etc., probably have more in common than not. The evolutionary case for veganism, which ties together a number of issues evident in this book, could be a point of common ground where fringe groups make contact. On the perimeter, groups like these listed, could engage in open debate to accommodate other viewpoints in an effort to cohere, address the opposition, and achieve real policy or cultural change.

Cooperation is the answer to solving world health, hunger, and climate issues. We already have some collaboration but require much more collective action. The United Nations has been working on these problems. While what follows might appear as an Us v. Them stance, that's not the case. Those postures regrettably exist at this time and will persist in the future, no doubt, as omnivores denigrate vegans and vice versa. Rather, I'm hopeful that my writing will help vegan skeptics realize that if we can gather around the evolutionary biology of hominins and the cultural evolution theory of social norms, we stand a chance to tackle some of the greatest challenges facing humans. The aim is not to convince you to become vegan, but to convince you that a vegan economy with vegan agriculture is a large part of the solution to our manifold global problems of worsening health and climate change. If I can do that, then maybe you will be the one to persuade someone else, who is almost there, to become vegan, creating a domino effect.

What Type of Tomorrow Do We Want?

One objection to my line of argument might be that I am falling into the fallacy of trying to devise an *ought* from an *is*. A complaint might run like this: just because our ancestral relatives were mostly plant eaters, like our extant cousins, great apes, that does not equate to who we are or should be. Putting aside accusations against this logic, the real concern is about the type of tomorrow, next week, or future we want. It's not just about what we are *capable* of accomplishing, but what we *should* do. Looking at the diet of our hominin past and relation to great apes and hunter-gatherers could help turn the lens a bit so that we see for sure that individual and environmental health spring from understanding our past. Each of us has the ability to look at the historical, scientific record to make an ethical choice about what we must do collectively for the good of ourselves, our children, and future generations.

While there is no deliberate effort here to directly address the United Nations (2020) sustainability goals, basic awareness of them is fundamental for anyone concerned about the longevity of nature and its plentiful plant and animal species. Population-wide changes in farming and diets leaning toward veganism are necessary for inextricably linked human and climate health (Fresán & Sabaté 2019; Bruno, et al. 2019). To that end, here are the U.N. objectives that are touched on, at least in part, and which could be kept in mind as one reads. Indeed, the leaders of the U.N. seem to be thinking that if an ecologically minded trend is absent or broken, then a correction should stem from a socially moral response. Pathways to achieving these targets need not be completely technological. Educating young people, altering attitudes, and adjusting values and beliefs can be prime movers as well.

U.N. goal 2: food security and sustainable agriculture come up in this book variously, mostly in envisioning a shift from industrialized meat manufacture to organic, local plant farming. Goal 3: the health of all people is tackled directly since most corporate food systems in developed countries, at times spilling into developing economies, are causing disease and bad health. Goal 6: sustainable water availability and waste processing are both implied in numbers 2 and 3 because animal farming generates much waste. Goal 9: infrastructure innovation is mentioned since we have the ability to refabricate existing meat processing plants to vegan food producers. Goal 11: sustainability in urban areas is imagined if suburban farms are established to feed cities, supplemented with urban community gardens. Goal 12: sustainable product consumption and production patterns are concerns implied in many other goals since the research, as you will read, says that animal farming is frequently unsustainable and unhealthy. Goal 13: action against climate change is one of the main areas addressed in this book because research shows how adopting a vegan culture will minimize and reverse the damages of climate change. Goal 14: managing ocean resources sustainably is implied in animal farming since that includes any number of seafood meats that are harvested with little regard for biodiversity conservation or pollution abatement. Goal 15: ecosystem and biodiversity protection and reclamation arise in parts of the discussion that address rewilding, which is attainable and can help sequester carbon dioxide.

Are We Ecological Inhabitants of Earth?

In a special issue of the *Ecological Citizen* (2020), in spite of its admirable focus on the significance of an eco-centric vision, there was no discussion of veganism. In earlier issues up to that point, there was one short opinion piece on veganism. Likewise, in a book entitled *Sustainable Modernity* (Witoszek 2018) there's no discussion of veganism. Sustainability means avoiding over consumption so that resources can be replaced or renewed, and we are not practicing that now. Pim Martens (2020) says that animal well-being, both domestic and wild, is a key component to sustainability. In our long evolutionary history, most of which has been as hunter-gatherers, we've coevolved with animals as coworkers (horses) or as pet workers (dogs). As part of E.O. Wilson's (1984) biophilia hypothesis we bond with other species, allowing for cultural differences. An ecological citizen with a concern for sustainable development would advocate for a vegan economy. I say vegan economy (or veganism or vegan agriculture or vegan culture) since there is no longer a portrait labeled "man the hunter."

The man the hunter theory minimizes gathering, foraging, and early, modest forms of agriculture to claim that everything human, from intellect to social life, derives from a hunting adaptation, the "master behavior" (Lee & DeVore 1968, Chapters 32 and 33). This faulty assumption persists in various forms across cultures, and it's the main theory I will attempt to dismantle. In his book *Brutal*, for instance, Brian Luke (2007) says there's masculine exploitation through justification by making animals purposeful for humans, whether as edible meat, objects of experimentation, or pets. He definitively states that men choose to exploit animals routinely, whether in hunting or for entertainment, and points out how surveys and studies consistently show women as more prone to animal rights activism than men. Many women eat meat perhaps because of cultural or class conditioning in a patriarchal society.

Let's be clear from the beginning about some key words. This might not be repeated, so pay heed. Dairy means any milk coming from a mammal, produced for its own offspring, but processed in such a way that it becomes human food as butter, cream, cheese, etc. Though not dairy, eggs are animal protein. When the word meat is used, I mean animal flesh, including mammals, seafood, and birds, but I'm not including insects in that definition. Julie Lesnik (2018) highlights the role of insect eating in the hominin diet, foraged mostly by females, although entomophagy offered an occasional, supplemental food source. The term meat need not refer to animal flesh, as some could say let's eat the "meat" of a nut. In a like way, Brianne Donaldson and Christopher Carter (2016) talk about the future of meat without animals. When any form of the word vegan is used, I am not referring to the now popular, highly processed animal look-alike foods. By vegan is meant raw or cooked vegetables; fruits; peas, beans, seeds, nuts; and negligibly processed meat and cheese substitutes consumed moderately. Foods like cereals and breads are allowed, if marginally processed using whole grains. Similarly, low processed brown, red, or black rice is nutritious.

When talking about a vegan economy, I do not mean multinational corporations that will spend billions of dollars researching, developing, and promoting faux meats or lab meats. For the past fifty years or so any social responsibility of business, it seems, has been to increase profits (Friedman 1970). In fact, economist and public policy analyst Jeffrey Sachs (2021) offers an alternative to free-market, profit-driven economics education in a revision more socially and environmentally ethical. I'm thinking of local farms and distributors, and small, regional businesses that produce meat substitute foods minimally processed. Such organic farms and producers would utilize natural power, like solar or water, for up to 30 percent of their energy needs. These suburban vegetable farms would be outside but near the borders of a city, supplemented by community gardens within the city, and perhaps hydroponic indoor farms on a limited capacity. Hydroponic farms use no dirt or pesticides, less water than conventional farming, but tend to require more attention, technology, and electricity than otherwise. The producers of plant-based foods would inhabit unused restaurants or abandoned commercial kitchens. Many municipalities already engage farmer markets. In the scenario being set, the plant foods would also be transported locally, cutting down emission pollutions, to city producers crafting vegan foods that can be distributed in neighborhoods.

Some communities, whether urban or not, don't have access to healthy foods. A. Breeze Harper (2010) points out how healthy diets, especially for Black or Brown people, can be limited by food impoverished locales or by institutions and culture. A challenge is to help people of color realize that veganism is not quite part of a privileged, white class but can be embraced by and benefit all people, especially those who are historically marginalized by racism, sexism, or colonization. In a nutshell, here are guideposts. This dilemma could be mitigated by coordinating an influx of vegetables from local farms, establishing community gardens and contingent plant-food workshops, and urging local leaders to establish vegan markets in areas where fresh veggies are typically unavailable. Church groups, school leaders, and even law enforcement could become part of the proposal. Obviously some community outreach, planning, and networking would be involved in establishing this plan, which is outside of the scope of the argument presented here.

In part, the solution to address neighborhoods with impoverished vegetarian food supplies relies on a combination of choice, public awareness, and willpower. Young people engaged in this effort would learn the meaning of ecological citizenry. Community green gardens are not new, nor are farming communities, but both seem to have languished over the past century. As a new model, one could turn to places like Serenbe, Georgia in the U.S. where a housing development is built around a small, working farm. This type of agrihood could become part of regular suburban, permaculture planning, land management employing systems thinking to mimic the synergies in nature supplying food and energy while reusing waste. Moreover, beyond the gratuitous "green spaces" in urban planning, pocket farm neighborhoods with cooking facilities for the manufacture of vegan foods could gain prominence in a city landscape.

Ecological citizens could lobby local elected officials for this progressive transformation. Existing concrete spaces or sections of city parks could be renovated. The focus in a vegan economy at the level of urban gardens could be on the healthy benefits of growing legumes, which do not require nitrogen fertilizers and promote biodiversity in their plots below and above the soil (Open Access Government 2020). Members of a community could lobby elected officials to help in this cost-effective effort. Contrary to what most people who go to a butcher shop or supermarket believe, meat is not "fresh." It's a dead animal. If read closely, prominent science researchers like Peter Andrews and R.J. Johnson (2019) imply we are evolving adaptations to digest grocery store foods. Consumption of any kind of meat three times a week or more leads to non-cancerous outcomes like debilitated circulation and respiration as well as digestive disease; red meat is responsible for higher saturated fat levels and colorectal cancer (Papier, et al. 2021).

Some fresh foods can be grown without a garden. Lettuce can be grown indoors. Microgreens are vegetables and herbs not fully mature but which have higher concentrations of nutrients than the adult plants (Xiao, et al. 2012). These can be grown at home or inside schools to complement schoolyard gardens. None of this is breaking news. Mid-nineteenth-century British publications like the *Vegetarian Advocate* and the *Vegetarian Messenger* extol the improvement of physical and mental health as well as moral behavior through plant foods. In his *Journal of Researches* (1839, Chapter 8), Charles Darwin talks of the "economy of nature" regarding a balance of species to food and (Chapter 20) an "organic arrangement." However, the advantage rises for those who, according to Darwin (1859, Chapter 4), capitalize from new economies. In this argument, that new economy is veganism if we decide to eliminate the damages of animal agriculture. Recognizing the centrality of plants, Karl Niklas (2016) says the evolution of photosynthesis literally brought breath and food to all other organisms and continues to do so. The primitive masculinity of many men is threatened by existence without red meat, and that perceived need is capitalized upon by advertisers (Rogers 2008). Carol Adams (2010) reflects on the publication of her book *The Sexual Politics of Meat* twenty years after its release and notes that there has been resistance with efforts to uphold "manhood." She rightly says that veganism is more about consumption than production and, therefore, it's really a boycott of animal agriculture. This is not to say we can't produce wholesome vegan foods. Of course, even with non-meat products that cause environmental degradation and exploit workers (like bees) one could argue for a boycott, as in the U.S. almond industry (Armstrong 2020).

Veganism represents knowledge about the cultural ecology of food with the implied question of what will be one's next, ethical move. More recently, scholars view veganism as Western "political consumerism" encompassing an array of social media campaigning, awareness blogging, and reforming events. Plant foods are often now marketed as "cool." Contrary to my contention, these scholars (Jallinoja, et al. 2019) claim that veganism is not a subculture but mainstream because of celebrity endorsements presented as an "inclusive movement." That's a question of what's in vogue, and those patterns, opposed to ethical veganism, could erode quickly. While their data indicate an upward trend of vegetarians and vegans worldwide, indisputably the meat and dairy producers hold a huge market share. Though mine is not a political argument per se, it's difficult to avoid the politics or ethics of animal farming, and there could be among the largest meat consuming societies a more robust public, activist discussion and rejection of the ills surrounding meat and dairy agriculture. Piia Jallinoja, et al. (2019) suggest as much and, therefore, substantiate my argument and claims. People want to be healthy and raise their children to thrive in a flourishing environment. Vegan organizers could capitalize on those assumptions.

Since Carol Adams (2010), scholars are looking at more nuanced entities within macro-cultures. Men and meat are culturally combined in various ways, not as a simple manly type, to express masculine identity differently based on roles from socioeconomic class. Julie-Anne Carroll, et al. (2019) suggest that meat plays a key function in the lives of men regarding social interactions, conspicuous consumption, and independence in some traditional and yet more evolving patterns. For reasons of health and environment, we should cultivate the embryonic micro-cultures that are moving away from heavy preferences for meat and dairy. With a shift in perceptions, most contemporary human communities could survive well on a vegan economy. Virtually all modern societies manufacture meat and dairy products behind public view and sell them neatly packaged in supermarkets. Farming and animal agriculture go back, in some places, to about 10kya, and while forms of hunting for subsistence persisted, most societies, not counting hunter-gatherers, were not engaged in a hunting economy. Early humans, like extant great apes, ate meat opportunistically, not religiously. Hunting for meat was not a cultural driver in our ancient ancestors.

Education, Awareness, and Influence

This is a brief but in many ways an overarching section for the book. Animal ethics taught in a classroom could have implications in the real world. This proposition was addressed by Eric Schwitzgebel, Bradford Cokelet, and Peter Singer (2020). These authors say that young people can experience behavioral change and overcome ethical dissonance toward animals. The U.S. study was based on a single class meeting of four very large groups and their attitudes about eating meat. Half the students read about animal farming, watched a video, and engaged in a discussion about animal ethics. The control group of half worked more broadly on the notion of charity. Then, the researchers examined for the entire semester meal card purchases totaling well over 10,000 receipts for about 500 students. The prediction was that the animal ethics class and discussion would have no effect.

However, the results indicate that for those students who discussed the moral complexities of farming and eating animals meal purchases for meat declined by 7 percent and remained stable for weeks. There was no change in meat eating in the group that discussed charitable giving. The researchers consider any drop in meat purchasing sustained over even a short time from one class meeting quite significant. There were variables. The authors of the paper say that in such large groups there was probably social or emotional influence, where experiences of those who already ate less meat demonstrated how normal that behavior really is. In line with the argument of this book, the authors go on to say that logos-based rational claims might have exerted important guidance in any student's ethical decision to eat less meat. Young people need options, not dictates. Juvenile primates, like gorillas and Japanese macaques, says George Schaller (1964), appear more receptive to new foods in contrast to adults. A positive outlook is to offer a vegan ethos so individuals young and old can decide on and adapt to the diet that fits their conscience.

This paper (Schwitzgebel, et al. 2020) supports claims about the importance of informing young people, and more broadly any given community, about the reasonable claims of avoiding meat and dairy products for personal and environmental health, to say nothing of the implications concerning animal ethics. Outside of the classroom, we need to get and then hold the attention of young people to these issues. The lesson of making moral choices from brief instruction should be kept in mind as you read this book. The evolutionary case for veganism is about shifts in cultural attitudes, values, and beliefs that are advantageous to humans, ecosystems, and animals. Educators and community leaders need to be more forthcoming in sensitizing people about the overall benefits of a vegan culture. If food is culture, and if moral norms inhere in culture, then there is an ethic about food and eating. Cultural leaders have an obligation to teach a food ethos that centers on the health of a person in a larger, ecological environment. The authors of the classroom ethics report confirm their findings in a follow-up empirical study (Schwitzgebel, et al. 2021) attesting how education can be a powerful influencer on the unidirectional behavior of young people in terms of making positive changes to their health, the global climate, or animal lives. Cultural contexts could heighten or lessen the effects of such educational influence.

Other scholars, too, conclude that meat eaters dissociate their empathic and disgust emotions from the reality of what's put in their mouths (Kunst & Hohle 2016). This dissonance occurs in how the meat is presented (e.g., a dismembered and headless corpse elicits less revulsion) or the language used (e.g., free-range rather than slaughtered or beef rather than cow). Such stilted attitudes against living creatures arise from culture and education. This disconnection of basic sympathy for animals yet killing them for food is a meat-animal paradox (Dowsett, et al. 2018). The appetite for meat stems from categorizing animals as food without moral standing, as objects without feelings, and as inanimate things that don't suffer (Bratanova, et al. 2011). Those distorted perceptions are cultural values, but they could be altered through educational awareness. Researchers also show that we are not purely objective thinkers; instead, our beliefs can determine how we see the world. Specifically, studies indicate that teaching faulty ideas regarding how an animal is raised, like the happy farm critter from children's books, determine affective responses to the meat on one's plate (Anderson & Barrett 2016).

As finishing revisions were made to the core of this book, I discovered Isabel Rimanoczy (2021) and her principles about a mindset of sustainability. Remarkably, there are some parallels to my thinking that readers should remember as the following pages unfold. For instance, Rimanoczy talks about an ecological worldview that is often narrow since it matches one's own interests. In contrast, one should become more broadly aware, eco-literate, and in touch with personal feelings as they connect to others. There's also a systems perspective that should move away from linear thinking to something more open. Seeing the world as set one way is not sustainable. Instead, relationships, patterns, flows, and connections need to be seen by looking at a big picture in the long term. Emotional intelligence also factors into Rimanoczy's sustainability mindset. Self-awareness is not just adaptive, rather it is preventative. One needs to question socially accepted values such as achieving wealth and material status over conservation. Here, she suggests not becoming unreactive and resigned to fate. Rather, innovation and adaptability in reflective consciousness are crucial. Finally, there's what she calls a spiritual intelligence. In opposition to what's simply utilitarian or rational, sustainable thinking should veer away from possessive consumption and self-centeredness to an intuitive oneness with nature to nurture compassion and empathy. Mindfulness is a practice with results; an embodied experience.

What's notable is how symbiotic mentality is precisely on point with my message: ecological food changes need not come solely from technology. Attention should shift to moral attitudes emerging from education, not just in history and science but in philosophy and particularly the arts to transform individuals fundamentally by breaking tightly held conservative preconceptions about health, the environment, and animals to embrace a more liberal world vision. Most vegans were raised as meat-eating omnivores and successfully adapted their awareness to a more sustainable way of thinking, so others can, too.

By the same token, Zoe Weil (2016) argues that we need to educate from an early age what she calls solutionary persons, not competitors, or those who work humanely and sustainably to root out and correct "exploitive systems" like animal farming. How one lives ethically is not just about a community of people but also related to the ecology of food, celebrating without harming the diversity of animal life. While skills in education are important, how one lives her life as a moral individual is paramount. We are not fully educating children to be friends of the natural world, animals, ecosystems, and less advantaged people. Weil discusses competencies around advocacy, creativity, ethics, skills transfer, sustainability, and collaboration, all of which can be applied to veganism. Education should not be separated into compartments that never intersect. Understanding the historical past is useful in developing new contexts, part of my interdisciplinary argument. As Weil says, we need to help people, especially youngsters, grasp what is true and not false. For example, veganism is not a cult, and climate change is not a conspiracy theory. Animal-based food culture is detrimental to health, the environment, and nonhuman sentient and sapient creatures. Confronting and not ignoring problems can lead to solutions, like sustainable choices changing the food system fundamentally and beneficially for all living beings.

Finally, driving home the central point of this book about the importance of education, look at David Nibert (2002), who sees animal and human rights as connected since, from a sociological perspective, there are many systemic economic, cultural, and political ideologies that foster violence and oppression against those who are vulnerable. He suggests that less culturally sanctioned exploitation of animals for economic gain could reduce violence against defenseless people, too. Putting logic to a test, ethics philosopher Nick Zangwill (2021) argues that humans have a moral obligation to eat meat. More reasonably, scientists Lori Marino and Michael Mountain (2021) explain how human killing of animals is an attempt to avoid confronting one's mortality. The devaluation of animal life is ingrained into many institutions because they are the beneficiaries. Reversing this course in educational institutions can be a positive step.

Eating Animals

While it's true in our human ancestry early hominins were fruit and leaf eaters (Chapter 4), the "evolutionary" part of my title refers to cultural evolution (Chapter 5). This means that because of our mental flexibility and technical ingenuity we can evolve away from reliance on eating meat and dairy. For some time, researchers have been able to make "animal" proteins from microorganisms without touching animals in what's called an acellular process. We are not quite evolutionarily programmed to be vegans, but we can make that ethical choice for our personal health and sustainable fitness of the planet. I am not pushing a radical vegan agenda. Instead, I'm thinking more along the lines of a public awareness campaign for significant movements away from meat and dairy farming. If we can accomplish this through cultural evolution, people will be making their own decisions and, therefore, no individual's liberty will be abrogated or taxed. For my argument, some will say there's no way most modern, industrial societies can shift to veganism. Maybe not, but we have to make an effort when considering individual health, the longevity of the environment, and the fate of animals. This is not a radical proposal. Besides, the call is for entrepreneurs and municipal legislators to foster vegan economies locally. Imagine a society where almost every school has a vegetable garden and young people are reared to revere their health in a sustainable ecosystem of food.

If there's a schoolyard garden, presumably with a kitchen in the school, or a community garden, perhaps with a kitchen in a nearby church or community center, vegan foods could be prepared. Vegan stir fry, patties, fritters, tacos, loafs, salads, or wraps are easy to make. Young people can gravitate away from junk food to become self-sufficient. See, for instance, the Dyckman Farmhouse (2021) and kitchen lab, Inwood, Manhattan, which teaches children and adults about green gardening, supplies grow kits, and offers vegetarian and even vegan recipes. Later, the national economy might catch the spirit of such positive change. Here's another example. Organizations like Grow NYC and Earth Matter support community gardening. What was originally established as a teaching garden a few years ago on Governors Island, between lower Manhattan and Brooklyn, converted into a fresh food supply source during the 2020 pandemic. Given the infrastructure on that island, which was once a military operations base, kitchens could prepare vegan foods from the abundance of vegetables and fruits grown. Some have proposed a casino at this location. Imagine the social benefits of a community garden and kitchen operated by local entities governed by minorities and women provided with city, state, or federal grants. Then, magnify that model across a city, state, country or globe with other decommissioned public institutions large enough to accommodate plant gardens and kitchens. Think how all of the Malls across the industrialized world that have been decommissioned could be converted to indoor veggie and fruit farms, where former restaurants there could be established as vegan kitchens. With the right funding, much of this food could be donated to shelters or given to needy families.

Most of the bad effects of meat and dairy are happening on a local level anyway. Factory farms won't shut overnight but could be phased out and repurposed. Many agricultural creatures need not be born into a life of cruelty where their confinement and persecution contribute to ecological degradation. Vegan agriculture uses far less land than animal farming, is environmentally optimal, and can feed more people in a nutritional manner with far less energy loss and waste (Shepon, et al. 2018; Eshel, et al. 2019). In 2006 the United Nations Food and Agricultural Organization released a report indicating that 18 percent of greenhouse gas emissions come from industrial animal agriculture, noting a "community danger" (Bristow & Fitzgerald 2011). That percentage is more than pollutants from worldwide transportation. By 2010 global greenhouse gas emissions from agriculture, forestry, and other land use jumped to 24 percent (Edenhofer, et al. 2014). The Böll Foundation (2014) says meat and dairy farming are responsible for upwards of 34 percent of global heating through direct and indirect fouling emissions. Environmental researchers Matthew Hayek and Scot Miller (2021) say greenhouse gas emissions from concentrated animal farming in the U.S. have been underestimated and compromise sustainability projections, taking into account, furthermore, expected increases in global meat consumption. The U.N. Food and Agricultural Organization estimates of emissions from animal agriculture are out of date, ecologically inefficient, and should be revised upward (Twine 2021). In whichever year you are reading this book, you see the upward trend line.

In light of these staggering statistics, certainly higher by now as rain forests in the Amazon are razed to breed massive herds of cattle for beef hamburgers, there's been little public discourse of this subject. Meat and dairy agriculture have been perpetuated through media efforts to rationalize the risk. Advertising is big money. One resolution is to offer an alternative to personal ill health and environmental catastrophe in veganism, which can begin with individual choice, spread in a family, and grow in a community. In urban areas of industrialized societies, many people are ignorant of how meat is produced. They only see sanitized packets of raw flesh in markets where the reality of the source has been erased. Furthermore, the meat industry tries to make it seem as if this antibiotic and hormone infused flesh is naturally occurring. Vegan chef and entrepreneur Miyoko Schinner (2020) tells the story of Oliver the pig, rescued from a meat plant and cared for in her sanctuary. Oliver was injected with growth hormones and antibiotics at birth and within a year became 700 pounds, only to die from a septic knee that could not hold his weight. Ironically or not, a modernized society usually has more educated and open-minded citizens, but yet their beliefs and attitudes about protein and nutrition are mistakenly tied to meat, fish, eggs, and dairy, an antiquated outlook on life. C. Eddie Palmer and Craig Forsyth (1992) note that there has been a cultural shift over time, and they write approximately thirty years before me. In most developed societies today, animals are rarely seen as beasts of burden or meat sources from the small family farm. Attitudes have expanded from welfare concerns and animal cruelty laws to animal rights. The next step is toward wider veganism, which these authors actually predicted through their data study and analysis in 1992.

The attempt in composing this work is not to solidify existing intellectual or animal-friendly relationships, but to challenge us with a daring perspective percolating in some circles: why can't human diets consist of virtually no animal flesh or dairy products. Scientists Lesley Newson and Peter Richerson (2021) claim that there's no real, natural human diet. It's cultural since the same protein, nutrients, and vitamins come from a variety of plants and animal foods. Plant-based diets are richer in protein with less risk of mortality (Abete, et al. 2014; Dagfinn, et al. 2017). The point, to borrow a word from Newson and Richerson, is that we now need to *renovate* the diet we inherited from our recent ancestors and reform it into something healthy and environmentally friendly. That can be done with some effort and a little imagination, evidenced in the many existing vegan groups and cultures.

How Culture Shapes Perceptions

Academics have some obligation to engage publicly in debates about social issues. Andrew Hoffman (2015) addresses that assertion in terms of climate change. I ask the same about veganism and all the culture wars springing from this misunderstood "lifestyle." In part, Hoffman says, ignorance and avoidance seem to rule over statistics. A *Lancet* commission on healthy eating and sustainability (Willett, et al. 2019) recommends a diet of mostly vegetables, fruits, whole grains, legumes, and nuts, with low consumption of poultry or seafood and near zero red meat. Not only, says the commission, will such a diet promote global health, but it would also reduce environmental degradation since food production is a key driver of climate change. Poor farming techniques on any scale endanger the health of humans, localities, and the planet's climate. Threats to biodiversity and land loss can easily be mitigated by focusing on the most affected agricultural areas, swiveling the human diet away from meat and dairy to increasing plant food yields, and eliminating food waste (Williams, et al. 2020). Academic research concerning food ecology does not often become mainstream news.

Without pressure from average people and academics, industrial agricultural polluters might not bow to social pressures about their role in and responsibility for climate change. Of nearly three dozen large meat and dairy companies, only four have committed to net-zero emissions by 2050, not by reducing animal methane gas emissions but by mitigating energy use (Lazarus, et al. 2021). This sounds like business as usual to breed and raise for slaughter as much livestock as possible. In addition to high levels of CO_2 now in the atmosphere, IPCC (2021) identifies methane (CH_4) as equally deadly to the climate. Much methane comes from massive animal agriculture via livestock breeding and is disrupting natural equilibrium in the atmosphere. Since methane does not have as long a life as CO_2, it can be reduced expeditiously in a transfer toward a vegan culture. Energy conversion or conservation is not the only means of reducing greenhouse gases. Replacing livestock production on huge tracts of lands with reforestation or in massive factory farms with plant-based alternatives can cause beneficial climate and health effects sooner rather than later. In many industrialized cultures and developing countries, information like that could be difficult to sell.

Cognitive filters or dissonance, not scientists, shape public debate based more on ideology or cultural identity and not on data. By analogy, analysis suggests that because of complex cultural values regarding facts, voters tolerate rather than punish, in spite of potential harms, known corruption in candidates seeking political office (De Vries & Solaz 2017). Many status quo economic and political interests are threatened by a vegan economy, and so obfuscate the issues laid out about health, the environment, and animals. Hoffman (2015) says this is true of the climate change debate, too. While I'd like the debate about veganism to be about health, the environment, and animals, public opinion might not have it that way. Rather, tribal schisms are typically set by a range of ideological, political, and corporate power brokers, not educators. That dynamic could change with some effort.

Melded into veganism are substantive issues about climate change. One study says that even among animal products, those of the lowest impact on terrestrial and aquatic ecosystems far exceed vegan substitutes (Poore & Nemecek 2018). Moreover, though I won't delve into it, veganism includes an evaluative issue about life. Some people who oppose abortion are "pro-life," yet they heartily consume pigs, cows, chickens, turkeys, lambs, fish, and other animal flesh. People and livestock animals for human consumption far exceed the biomass of wild animals (Goodland & Anhang 2009; Bar-On, et al. 2018), and this imbalance contributes to species extinction and environmental catastrophe. Citizens, whether global or regional, need to be informed by authoritative information like this. Because of human intrusion and activity in many parts of the world there is no longer abundant earth. There is scientific consensus about veganism regarding its benefits for health and the environment (e.g., De Boo & Knight 2020), but tribal mentalities will suppress any such agreement. Steaks, after all, rule the food economy, not tofu.

Scientists worldwide have dire predictions about climate change based on 14,000 studies (IPCC 2021), yet some leaders and many people will question the findings and warnings. As Hoffman (2015) might say, the scientific consensus is meaningless if there is no social unanimity. A goal is to help achieve accord about veganism through cultural evolution, which could construct a dramatic shift in ideology. As with those who question vaccines (Horne, et al. 2015), the emphasis in an argument for veganism should highlight the risks to health and mortality, environment, and animals by not choosing plant-based foods. Countering misinformation and false beliefs is accomplished not so much with a mountain of factual evidence showing why veganism is beneficial but through social psychology. Listening to and communicating directly with non-vegans to foster "intergroup contact" is essential in overcoming bias (Pettigrew, et al. 2006). With some work, we could establish a grid for a vegan economy. Reaching out to and communicating with young people is an essential element here since they hold the keys to the future of their health and the planet.

Policy decisions are not made in a political or cultural vacuum. Bad policy that harms health should not continue only because a majority of the public chooses the poisonous over the salutary. I understand how one's reasons and motivations are formed by cultural cognition or group values. My writing is a clear manifestation of that. People choose sides and don't want to change their minds, much less have someone else change their minds for them. Since science is investigation, discovery, and theory so that values can be modified or altered in a meaningful and significant way for all to prosper, we should be willing to alter our beliefs and attitudes. Think, for instance, of Copernicus and Galileo or Newton and Darwin. We should be motivated to ethical action by facts and not by alternative facts. We can, certainly, be ethically motivated by emotions and feelings since we are not, thankfully, robots. There's the difficulty, but it's also a blessing. Anyway, prevarication would loom if one says this appeal has no reference to shifting moral norms.

As a sociologist might say, people take hard facts about veganism and distort them with their tribal values about religion, identity, family, food, etc. Perhaps I'm doing that. Liberal thinkers can be discriminatory about others' values, though less severely than conservatives (Wetherell, et al. 2013). We tend to limit our thinking to what is personally important. In terms of climate change, Hoffman (2015) asks a few applicable questions. Whom do you trust? Do big corporations care about social health issues or global overheating? Do you, reader, trust me, the scientific process used to create this message and the solution offered? Even within the camp I'm supporting there are divisions, and let's name only two. Some are ethical vegans who foster mental and physical well-being, care for the environment, and avoidance of harm to animals. Others see veganism as a means to profit. In fact, I do rhetorically appeal to that second group by, sometimes, using the term vegan economy. My use of "economy" is, however, local, small, environmentally friendly, and ethical toward people and animals.

At any rate, the two sides of this vegan economy I envision speak completely different languages. On the one hand there are virtue ethics, on the other there are financial gains and losses. The food industry is big business, and multinational corporations and venture capitalists are already staking claims in their version, not the vision, of a vegan economy. One word everyone understands is risk. By continuing with an economy of meat and dairy farming we are risking much more loss overall than if we migrated to vegan agriculture. Rather than, at first, choosing sides, we should develop a consensus. That's the goal in laying out the facts of biological and cultural evolution. This writing could be a community builder, though I'm not naïve and understand that hardline meat eaters will shout a resounding No and not consent. From the early 1900s to now U.S. Americans have nearly doubled their intake of meat and dairy products per capita (Stone 2011; Ritchie & Roser 2019). Hoffman (2015) drives home the point that any debate is about values and not science. We need to consider to what degree people are encultured to eating meat and dairy rather than making informed dietary decisions.

The views presented here, while gaining traction, are in the minority. Wealthy, industrialized societies might want to consider who determines what they eat, which then impacts health and the environment. There can be external influences to one's food purchases, whether parents, peers, or corporate advertisers. Any internet search reveals that the meat industry is waging war against those who produce and consume vegan products. One could imagine some of these massive meat vendors switching their factories from slaughterhouses to purveyors of plant-based products. Using a sustainability mindset, Isabel Rimanoczy (2021, Chapter 6) metaphorically talks about recycling reusable materials, opposed to harmful linear waste. Reinvention is part of the human psyche and at the core proposed in this argument. Elmhurst Dairy in N.Y. had been producing and distributing cow's milk since 1925, but by 2017 through visionary and imaginative leadership the company transitioned to plant-based "milks" and dropped the word dairy from its name.

Even better, rather than corporate control, meat executives could compartmentalize these huge complexes to smaller, individually (preferably minority) owned makers of marginally processed vegan meats, cheeses, and dairy products. Financially, this could be accomplished through profit sharing. During the 2020 pandemic there was a glut of animals, so these abattoirs recklessly killed or buried animals alive otherwise slated for someone's dinner plate. Oddly or not, this was the media story, and not how farmed meat and dairy lead to obesity and environmental ruin. In this book, obesity means excess body fat whereas overweight means excess body weight. If people don't read quality, peer-reviewed material, some media coverage can distort one's ability to reason properly. Young people should know that obesity will hinder cognitive and motor function over their lifespan (Wang, et al. 2016). One large observational study (Boonpor, et al. 2021) says people who eat more plant foods have healthier biomarkers or measures of disease than meat eaters. The body mass index scale (BMI) can be notoriously problematic, and there are overweight vegans.

Hoffman (2015) points out how there have been a host of events, like the publication of Rachel Carson's environmental impact book *Silent Spring* (1962) or the Santa Barbara oil spill (1969) or industrial disasters in India (1984) and Ukraine (1986), that punctuated immediate changes in social attitudes about life and health. Perhaps the 2020 pandemic, in spite of all the personal misery and economic destruction it wrought, could become another one of those turning points. Now seems to be the time to consider a vegan economy, but one that is less corporate-based and more local where megastructure animal trade is siphoned off to entrepreneurial individuals crafting healthy vegan plant foods in stores, restaurants, or mini workshops. If not through profit sharing, meat rendering factories can be refurbished as vegan franchises. For many people, climate change is still somewhat abstract, but veganism does not have to be since it deals with individual bodies. A few years ago, Hoffman said most Americans didn't think they would be affected by climate change, but recent catastrophes that have hit home might alter that thinking. Just as everyone is affected in some degree by climate change, everyone is surely the outgrowth of what he or she eats. Indeed, with less meat and dairy farming, much of the grains and water used to feed animals staged for slaughter can be shared with starving children.

For this argument, North Americans and many other people worldwide, based on obesity rates alone, are affected by eating too much meat and dairy improperly prepared. In N.Y.C., Brooklyn Borough President Eric Adams (2020) not only lost weight but beat diabetes, along with his octogenarian mother, by switching to a diet of plant foods. Such restoration of good health is not anecdotal but verified by other medical specialists advocating a vegan diet (Stone 2011). Eric Adams rightly says that food can be a form of slavery, keeping some populations in poor health and dependency. One study (Najjar, et al. 2018) shows how plant foods reduce cardiovascular disease and dependency on medications. An improper vegan diet can cause health problems, too, but that's not the purview of this writing and usually stems from user error. When I write some form of the word vegan I assume readers will understand I mean temperance and moderation, plant foods that are slightly processed, with lots of vegetables including some fruits, seeds, beans, and nuts. Beans offer an equivalent amount of protein found in cow meat but with substantially less health and climate risks. Moving to a healthy vegan economy that is not controlled by mega corporations selling their "burgers" solely for profit but that is spread across municipalities, states, and societies as part of the community fabric is a workable goal. Energy efficient suburban, urban, schoolyard, or vertical permaculture farms producing legumes, veggies, etc. could tackle a public health risk and ameliorate the devastating effects of climate change. While all of this should be a rational decision, for many it's emotional.

We need to carefully formulate the habits of mind we want to instill in younger generations. Culture drives human evolution more than biology (Waring & Wood 2021), but no transition is special if it generates problems from that culture. It's time for "culture" to be divided, as attempted in this book by contrasting, for instance, animal agriculture with a vegan culture. There are many forms of culture, some morphing more than others. Emphasis should be placed on those that will help humans survive and care for the planet. Rather than a purely capitalistic, dominant business culture, we should plan a future with an ecological, ethical vegan economy. Human gene survival depends on vegan education, learning, group formation, and influence for extended solutions. By now, you see the focus is on the gains to be made for all humans and animals if politicians, small businesses, and everyday people swing allegiance away from their polarizing ideology just a little and start to build a consensus since we all are facing common health and environmental problems. Extreme ideas are not invaluable. In fact, Hoffman (2015) notes how radical thinking can instigate some people to be reasonable, and I suppose that's the project, to generate communal willpower for positive change.

Let's move to the main chapters and get into discussions of scientific, philosophical, and cultural topics, offering opinion informed by authoritative and credible sources as we proceed. One key ingredient in this debate about food cultures deals with the type of nutritional community we envision for ourselves and descendants: ecologically sustainable.

References

Abete, Itziar, et al. 2014. "Association Between Total, Processed, Red and White Meat Consumption and All-cause, CVD and IHD Mortality: A Meta-analysis of Cohort Studies." *British Journal of Nutrition* 112 (5): 762–775. doi:10.1017/S000711451400124X.

Adams, Carol J. 2000. *The Sexual Politics of Meat: A Feminist-vegetarian Critical Theory.* Tenth Anniversary Edition. NY: Continuum.

Adams, Carol J. 2010. "Why Feminist-vegan Now?" *Feminism and Psychology* 20 (3): 302–317. doi:10.1177/0959353510368038.

Adams, Eric. 2020. *Healthy at Last: A Plant-based Approach to Preventing and Reversing Diabetes and Other Chronic Illnesses.* Carlsbad, CA: Hay House.

Anderson, Eric and Lisa Feldman Barrett. 2016. "Affective Beliefs Influence the Experience of Eating Meat." *Plos One* 11 (8). https://doi.org/10.1371/journal.pone.0160424.

Andrews, Peter and R.J. Johnson. 2019. "Evolutionary Basis for The Human Diet: Consequences for Human Health." *Journal of Internal Medicine* 287 (3): 226–237. doi:10.1111/joim.13011.

Armstrong, Lewis. 2020. "California's Almond Trade is Exploiting One of Nature's Most Essential Workers." *Thought for Food Blog.* International Food Information Service. https://www.ifis.org/blog/californias-almond-trade-exploiting-bee-population.

Bar-On, Yinon M., et al. 2018. "The Biomass Distribution on Earth." *PNAS* 115(25): 6506–6511. www.pnas.org/cgi/doi/10.1073/pnas.1711842115.

Böll Foundation. 2014. *Meat Atlas.* Second edition. Berlin: Böll Foundation.

Boonpor, Jirapitcha, et al. 2021. "Heath-related Biomarkers of Profile Vegetarians and Meat-eaters: A Cross-sectional Analysis of the UK Biobank Study." University of Glasgow Poster EP3–33. European Association for the Study of Obesity.

Bratanova, Boyka, et al. 2011. "The Effect of Categorization as Food on the Perceived Moral Standing of Animals." *Appetite* 57 (1): 193–196. https://doi.org/10.1016/j.appet.2011.04.020.

Bristow, Elizabeth and Amy J. Fitzgerald. 2011. "Global Climate Change and the Industrial Animal Agriculture Link: The Construction of Risk." *Society and Animals* 19: 205–224. doi:10.1163/156853011X578893.

Bruno, Morena, et al. 2019. "The Carbon Footprint of Danish Diets." *Climate Change* 156: 489–507. https://doi.org/10.1007/s10584-019-02508-4.

Carroll, Julie-Anne, et al. 2019. "Meat, Masculinity, and Health for the 'Typical Aussie Bloke': A Social Constructivist Analysis of Class, Gender, and Consumption." *American Journal of Men's Health* November–December: 1–12. doi:10.1177/1557988319885561.

Clark, Michael A., et al. 2020. "Global Food System Emissions Could Preclude Achieving the 1.5° and 2° C Climate Change Targets." *Science* 370 (6715): 705–708. doi:10.1126/science.aba7357.

Crary, Alice. 2018. "Ethics." *Critical Terms for Animal Studies.* Lori Gruen, ed. Chicago: U Chicago P. 154–168.

Dagfinn, Aune, et al. 2017. "Fruit and Vegetable Intake and the Risk of Cardiovascular Disease, Total Cancer and All-cause Mortality – A Systematic Review and Dose-response Meta-analysis of Prospective Studies." *International Journal of Epidemiology* 46 (3): 1029–1056. https://soi.org/10.1093/ije/dyw319.

Darwin, Charles. 1839. Journal of Researches. James A. Secord, ed., *Charles Darwin: Evolutionary Writings.* Oxford: OUP. 2008.

Darwin, Charles. 1859. *On the Origin of Species.* Joseph Carroll, ed. Ontario, CN: Broadview P. 2003.

De Boo, Jasmijn and Andrew Knight. 2020. *The Green Protein Report.* Auckland: Vegan Society.

De Vries, Catherine and Hector Solaz. 2017. "The Electoral Consequences of Corruption." *Annual Review of Political Science* 20: 391–408. https://doi.org/10.1146/annurev-polisci-052715-111917.

Di Paola, Ariana, et al. 2017. "Human Food v. Animal Feed Debate. A Thorough Analysis of Environmental Footprints." *Land Use Policy* 67: 652–659. https://doi.org/10.1016/j.landusepol.2017.06.017.

Donaldson, Brianne and Christopher Carter. 2016. *The Future of Meat Without Animals*. London: Rowman and Littlefield International.

Dowsett, Elisha, et al. 2018. "Neutralising the Meat Paradox: Cognitive Dissonance, Gender and Eating Animals." *Appetite* 123: 280–288. https://doi.org/10.1016/j.appet.2018.01.005.

Dyckman Farmhouse, Growing Uptown. 2021. https://dyckmanfarmhouse.org/growing-up town/.

Ecological Citizen. 2020. Patrick Curry, et al., eds. Volume 3 supplement C. www.ecologica lcitizen.net.

Edenhofer, Ottmar, et al. 2014. *Climate Change 2014: Mitigation of Climate Change*. NY: Cambridge UP.

Eshel, Gidon, et al. 2019. "Environmentally Optimal, Nutritionally Sound, Protein and Energy Conserving Plant Based Alternatives to U.S. Meat." *Scientific Reports* 9:10345. https://doi.org/10.1038/s41598-019-46590-1.

Fresán, Ujué and Joan Sabaté. 2019. "Vegetarian Diets: Planetary Health and Its Alignment With Human Health." *Advances in Nutrition* 10 (4): S380–S388. https://doi.org/10.1093/advances/nmz019.

Friedman, Milton. 1970. "The Social Responsibility of Business is to Increase Its Profits." *The New York Times* 13 September.

Goodland, Robert and Jeff Anhang. 2009. "Livestock and Climate Change." *World Watch*. https://awellfedworld.org/wp-content/uploads/Livestock-Climate-Change-Anhang-Goo dland.pdf.

Gruen, Lori, ed. 2011. *Ethics and Animals: An Introduction*. Cambridge: CUP.

Gruen, Lori and Kari Weil. 2012. "Animal Others." *Hypatia* 27 (3): 477–487.

Haeckel, Ernst. 1879. *The Evolution of Man*. NY: Appleton.

Handley, Carla and Sarah Matthew. 2020. "Human Large-scale Cooperation as a Product of Competition Between Cultural Groups." *Nature Communications* 11:702. https://doi.org/10.1038/s41467-020-14416-8.

Harper, A. Breeze, ed. 2010. *Sistah Vegan: Black Female Vegans Speak on Food, Identity, Health, and Society*. NY: Lantern.

Hayek, Matthew and Scot Miller. 2021. "Underestimates of Methane From Intensively Raised Animals Could Undermine Goals of Sustainable Development." *Environmental Research Letters* 16: 053006. https://doi.org/10.1038/1748-9326/ac02ef.

Henrich, Joseph. 2016. *The Secret of Our Success: How Culture is Driving Human Evolution, Domesticating Our Species, and Making Us Smarter*. Princeton: Princeton UP.

Hoffman, Andrew J. 2015. *How Culture Shapes the Climate Change Debate*. Stanford: Stanford UP.

Horne, Zachary, et al. 2015. "Countering Antivaccination Attitudes." *PNAS* 112 (33): 10321–10324. www.pnas.org/cgi/doi/10.1073/pnas.1504019112.

IPCC. 2018. *Global Warming of 1.5°C. An IPCC Special Report on the Impacts of Global Warming*. https://www.ipcc.ch/sr15/.

IPCC. 2021. *Climate Change 2021: The Physical Science Basis*. https://www.ipcc.ch/report/sixth-assessment-report-working-group-i/.

Jallinoja, Piia, et al. 2019. "Veganism and Plant-based Eating: Analysis of Interplay Between Discursive Strategies and Lifestyle Political Consumerism." *The Oxford Handbook of Political Consumerism*, Magnus Boström, et al., eds. Oxford: OUP. 157–180.

King, Barbara J. 2021. *Animals' Best Friends: Putting Compassion to Work for Animals in Captivity and in the Wild*. Chicago: U Chicago P.

Kunst, J.R. and S. Hohle. 2016. "Meat Eaters by Disassociation: How We Present, Prepare and Talk About Meat Increases Willingness to Eat Meat by Reducing Empathy and Disgust." *Appetite* 105: 758–774. https://doi.org/10.1016/j.appet.2016.07.009.

Lazarus, Oliver, et al. 2021. "The Climate Responsibilities of Industrial Meat and Dairy Producers." *Climate Change* 165: 30. https://doi.org/10.1007/s10584-021-03047-7.

Lee, Richard B. and Irven DeVore. 1968. *Man the Hunter*. Chicago: Aldine Publishing.

Leenaert, Tobias. 2017. *How to Create a Vegan World: A Pragmatic Approach*. NY: Lantern Books.

Lesnik, Julie J. 2018. *Edible Insects and Human Evolution*. Gainesville: U Press of Florida.

Luke, Brian. 2007. *Brutal: Manhood and the Exploitation of Animals*. Champaign, Ill.: U Illinois P.

Marino, Lori and Michael Mountain. 2021. "'I am Not an Animal': Denial of Death and the Relationship Between Humans and Other Animals." *The Kimmela Center*. https://www.kimmela.org/2021/11/26/i-am-not-an-animal-kimmela-white-paper/.

Martens, Pim. 2020. *Sustanimalism: A Sustainable Perspective on the Relationships Between Human and Non-human Animals*. De Biezen, Netherlands: Global Academic Press.

Najjar, Rami S., et al. 2018. "A Defined, Plant-based Diet Utilized in an Outpatient Cardiovascular Clinic Effectively Treats Hypercholesterolemia and hypertension and Reduces Medications." *Clinical Cardiology* 41 (3): 307–313. doi:10.1002/clc.22863.

Newson, Lesley and Peter Richerson. 2021. *A Story of Us: A New Look at Human Evolution*. Oxford: OUP.

Nibert, David. 2002. *Animal Rights/Human Rights: Entanglements of Oppression and Liberation*. Lanham, MD: Rowman and Littlefield.

Niklas, Karl J. 2016. *Plant Evolution: An Introduction to the History of Life*. Chicago: U Chicago P.

Open Access Government. 2020. "Agriculture Focus: Legumes Contribute to a Better World." https://www.openaccessgovernment.org/legumes/88777/.

Palmer, C. Eddie and Craig J. Forsyth. 1992. "Animals, Attitudes, and Anthropomorphic Sentiment: The Social Construction of Meat and Fur in Post Industrial Society." *International Review of Modern Sociology* 22: 29–44.

Papier, Keren, et al. 2021. "Meat Consumption and Risk of 25 Common Conditions: Outcome-wide Analyses in 457,000 Men and Women in the UK Biobank Study." *BMC Medicine* 19: 53. https://doi.org/10.1186/s12196-021-01922-9.

Pettigrew, T.F., et al. 2006. "A Meta-analytical Test of Intergroup Contact Theory." *Journal of Personality and Social Psychology* 90 (5): 751–783. https://doi.org/10.1037/0022-3514.90.5.751.

Poore, J. and T. Nemecek. 2018. "Reducing Food's Environmental Impacts Through Producers and Consumers." *Science* 360 (6392): 987–992. doi:10.1126/science.aaq0216.

Rimanoczy, Isabel. 2021. *The Sustainability Mindset Principles*. London: Routledge.

Ritchie, Hannah and Max Roser. 2019. "Meat and Dairy Production." *Our World in Data*. https://ourworldindata.org/meat-production.

Rogers, Richard A. 2008. "Beasts, Burgers, and Hummers: Meat and the Crisis of Masculinity in Contemporary Television Advertisements." *Environmental Communication* 2 (3): 281–301. doi:10.1080/17524030802390250.

Sachs, Jeffrey D. 2021. "From Neoclassical Economics to the Economy of Francesco." *Journal of Jesuit Business Ethics* 12 (1): 7–14.

Schaller, George B. 1964. *The Year of the Gorilla*. Chicago: U of Chicago P, 1988.

Schinner, Miyoko. 2020. *Rancho Compassion*. https://www.ranchocompasion.org/in-memoriam.

Schwitzgebel, Eric, et al. 2020. "Do Ethics Classes Influence Student Behavior? Case Study: Teaching the Ethics of Eating Meat." *Cognition*, 203: 104397. https://doi.org/10.1016/j.cognition.2020.104397.

Schwitzgebel, Eric, et al. 2021. "Students Eat Less Meat After Studying Meat Ethics." *Review of Philosophy and Psychology*. https://doi.org/10.1007/s13164-021-00583-0.

Shepon, Alon, et al. 2018. "The Opportunity Cost of Animal Based Diets Exceeds All Food Losses." *PNAS* 115 (15): 3804–3809. https://doi.org/10.1073/pnas.1713820115.

Springmann, Marco, et al. 2018. "Options for Keeping the Food System Within Environmental Limits." *Nature* 562: 519–525. https://doi.org/10.1038/s41586-018-0594-0.

Stone, Gene, ed. 2011. *Forks Over Knives: The Plant-based Way to Health*. NY: The Experiment.

Twine, Richard. 2021. "Emissions From Animal Agriculture – 16.5% is the New Minimum Figure." *Sustainability* 13 (11): 6276. https://doi.org/10.3390/su13116276.

Tyson, Alec and Brian Kennedy. 2020. *Two-Thirds of Americans Think Government Should Do More on Climate*. Pew Research Center.

United Nations. 2020. *The Sustainable Development Goals Report*.

Wang, Chuanming, et al. 2016. "Obesity Reduces Cognitive and Motor Functions Across the Lifespan." *Neural Plasticity* 2473081. http://dx.doi.org/10.1155/2016/2473081.

Waring, Timothy M. and Zachary T.Wood. 2021. "Long-term Gene-culture Coevolution and the Human Evolutionary Transition." *Proceedings of the Royal Society B* 288: 20210538. https://doi.org/10.1098/rspb.2021.0538.

Weil, Zoe. 2016. *The World Becomes What We Teach: Educating a Generation of Solutionaries*. NY: Lantern Books.

Wetherell, Geoffrey A., et al. 2013. "Discrimination Across the Ideological Divide." *Social Psychological and Personality Science* 4 (6): 658–667. https://doi.org/10.1177/1948550613476096.

Willett, Walter, et al. 2019. "Food in the Anthropocene." *Lancet* 393 (10170): 447–492. https://doi.org/10/1016/S0140-6736(18)31788-4.

Williams, David R., et al. 2020. "Proactive Conservation to Prevent Habitat Losses to Agricultural Expansion." *Nature Sustainability*. https://doi.org/10.1038/s41893-020-00656-5.

Wilson, Edward O. 1984. *Biophilia: The Human Bond With Other Species*. Cambridge, MA: Harvard UP.

Witoszek, Nina and Atle Midttun, eds. 2018. *Sustainable Modernity: The Nordic Model and Beyond*. London: Routledge.

Xiao, Zhenlei, et al. 2012. "Assessment of Vitamin and Carotenoid Concentrations of Emerging Food Products: Edible Microgreens." *Journal of Agricultural and Food Chemistry* 60 (31): 7644–7651. https://doi.org/10.1021/jf300459b.

Zangwill, Nick. 2021. "Our Moral Duty to Eat Meat." *Journal of the American Philosophical Association*: 295–311. doi:10.1017/apa.2020.21.

1

PRELIMINARIES AND OBJECTIONS

In this chapter, I hope to build on my reliability, presumably already established. Readers are reminded of the overall argument and key claims laid out at the beginning of this book dealing with a drive for community education and awareness about the health and environmental benefits of a vegan food culture. We are evolved apes who have mostly lost contact with natural biodiversity, the pattern of relationships in nature, as seen by Darwin. Pim Martens (2020) says that in line with biodiversity preservation, emphasis should be on wildlife and animal care, not simply on enumerating species.

Who is a Vegan?

Simply put, a vegan is someone who does not use or eat animal products, whether meat, seafood, or dairy, maintains an ethos that is free from any animal cruelty, and avoids purchasing leather, fur, products tested on animals, etc. Annie Potts and Philip Armstrong (2018) discuss how veganism can be a "marker of difference" where one politically resists any form of capitalism that profits from harming animals. As Barbara McDonald (2000) puts it, one might have an orientation toward being vegan-like, perhaps because of heightened empathy that gets set in motion by some catalyst. This is not to say others can't embody and act on the feeling of being vegan-like, evident in the argument and claims of this book. The invitation is open.

As stated in the Declaration of Toulon (2019), animals are persons and not things. A vegan can achieve health and fitness, evidenced in the many vegan athletes across a range of sports from weight lifting, running, and tennis. For the sake of my argument, when I use the term veganism I am not referring to raw veganism (Alvaro 2020a). With raw veganism one's diet consists almost exclusively of non-cooked foods like fresh fruit and leafy greens along with nuts or seeds. While that

DOI: 10.4324/9781003289814-2

could be an ultimate goal for some people, I don't foresee that as a practical strategy in winning over readers in this argument. Even some vegans will admit that the shift in their diet was a gradual weaning away from red meat, then away from chicken, and eventually away from dairy. Some cultures don't eat much meat since it's expensive or prohibited because of religious reasons. At any rate, and to emphasize a serious point from an evolutionary perspective, say Donna Hart and Robert Sussman (2005), up until about 10kya with the introduction of farming, meat was a "scarce supplement" to a variety of other, mostly plant, foods. I define the word vegetarian loosely to describe a person or animal who might rarely but not regularly be a consumer of some animal flesh or dairy, including the ingestion of insects. I will talk about eating insects in Chapters 3 and 4.

For many people, veganism means not eating any animal flesh and not consuming any dairy products. There are no steaks, chops, ribs, or fish. There are no animal-based eggs, milk, or butter. There is no honey. However, corporate executives have capitalized on a growing vegan trend, so there are many animal substitute products widely available, very tasty, and infused with vitamins and nutrients. One foreseeable problem is the corporate takeover of vegan agriculture, but I will not address that in this writing other than raising the red flag. Yet it's a concern, and any new vegan economy should be guided by wise policy makers and visionary politicians who keep the vegetable farms small and within a short distance from urban areas. Community leaders should allow entrepreneurial manufacture and distribution of veggie products, especially in city neighborhoods lacking healthy food. Suffice it to say that if vegan companies stay true to their mission, they could have multiple, energy efficient operations across the country to reduce the burning of fossil fuels in shipping. They could have controlled profit sharing or franchise opportunities for women and minority start-ups where vegan quality is maintained and improved along cultural food traditions.

It's probably futile to make some vegans feel uncomfortable because they don't eat raw foods or to make meat eaters feel guilty. Raising awareness so people can generate informed decisions is the goal. Meat and dairy eaters could be motivated to change course in seeing how there's an environmentally responsible, healthy, and ethical choice in veganism.

Identity Thinking

Vegans have a perceived and projected identity often rooted in an authentic ethos more than a diet (Greenebaum 2012). There's a clash between this minority vegan ethic, which strives to benefit the planet and its biodiversity, and consumers with little regard for the consequences of participating in animal agriculture. Part of the challenge, addressed in this book, is to use biology, evolutionary history, and cultural evolution as a means to rationally sway most people away from meat and dairy eating as supposedly integral to their survival identity.

Researchers (McCright & Dunlap 2011) recognize what is called the "white male effect." Conservative white men, apparently, are more likely than others in the United States to deny climate change across the board. Such monolithic attitudes, by analogy and since battling climate change is tied to veganism, are hard to overcome. This demographic (see, too, Pennycook, et al. 2019) perceives risk based on their worldview of rugged individuals in a hierarchical order. Such stolid thinking needs to be chipped away, and some closed attitudes might have opened up because of the pandemic and recent climate events.

Psychologist Joshua Greene (2013) would call group division moral tribes, but I wonder where the morality is in those who ignore violence in animal agribusiness and destruction in deforestation for cattle farms. Each group has its own set of values and beliefs, but not all *mores* are life-affirming or ecologically feasible in the long term. Visionary, eloquent leaders who appeal to such men (and women) could be a start, as well as a more well-informed medical community. When these men, and others for that matter, visit their doctors, there could be a discussion about the total health benefits of a vegan diet. Most physicians are not trained in nutrition, much less a vegan diet. The theory is that conservatives adhere to what's called identity protection and are reluctant, no matter how much data is thrown at them, to change their minds. If you can alter the thinking of someone close to such a man, then maybe that person could exert an influence. Furthermore, John Jost, et al. (2017) show how the conservative view is like a syndrome of anxiety about uncertainty, and so there's an alliance to the familiar status quo of traditions and hierarchy. Education (not indoctrination) of young people about healthy foods and the environment is paramount in a new cultural paradigm.

Confirmation Bias

We need to study evolution in light of growing pandemics and environmental collapse through climate change. Misunderstanding how scientific theory works, i. e., the testing of propositions, is a common problem. As Naomi Oreskes (2018) says, climate change is not a matter of a vote among a few scientists. Rather, via peer review, as a whole, large numbers of researchers work out key parts and questions of the puzzle. There is a consensus, but it gets refuted by a few often unscientific voices. Consensus correlates with expertise, as Cook, et al. (2016) demonstrate in a synthesis of studies indicating how nearly all climate scientists agree that the atmosphere and oceans are warming because of human activities. The farming of animals for meat and dairy is a human-induced factor (Di Paola, et al. 2017). Part of the problem involves making predictions about the future. That's not an obstacle with the argument for a vegan economy, for by reducing meat fat and dairy cholesterol we can improve health while reducing animal farm pollution.

You can be a vegan only for health reasons, but if so, the chances are that you have considered the ethical ramifications as well. We tend to interpret data to confirm our feelings or beliefs, even if those mental states are not substantiated by facts. Emphasizing education and a rise in science courses, as argued here, Jon Miller, et al. (2021) say that belief in human evolution is increasing. Nonetheless, Andrew Hoffman (2015) seems to suggest, when you hit a stone wall you are confronted by confirmation bias. No matter how much evidence you present to the contrary, the other person or group will shut down even more. Likewise, Keith Stanovich and Richard West (2007) describe how personal opinions become fixed in spite of factual evidence, although education could factor into one's critical thinking. Those with this so-called myside bias seem less affected by new information and more concerned about how new cultural ideas threaten them.

As many authors concerned with conservation biodiversity and environmental ethics point out, humans need not consume nearly as much animal flesh as they do. In fact, in many prosperous industrialized countries, from China to North America, people are ingesting too much protein. Ecologist Carl Safina (2011) points out how all "animals," from insects like bees to elephants, wolves, and dolphins exhibit a range of social intelligence and cross-species empathy demonstrating why we should not be killing them for convenience or food. More importantly, every living creature exists in symbiotic relation to another, thereby forming a stable and sustainable biome that contributes to what once was a self-regulating biosphere. Nothing could be further from the truth than asserting that humans are born meat eaters, programmed to hunt, or destined to eradicate other species into extinction, as I hope to demonstrate.

The American Heart Association (Kim, et al. 2019) says that overall health, and particularly cardiovascular fitness, is improved in a plant-based diet. Dr. Tushar Mehta and Nicholas Carter (2021) have compiled a comprehensive list of the best research on the benefits of a plant diet. Microbial gut bacteria promote good health and can prevent serious illness like diabetes and heart disease. A fit microbiome results from eating minimally processed whole foods low in salt, sugar, or other chemical additives. Not everyone's gut will react the same way, but researchers (Asnicar, et al. 2021) found that healthy gut bugs stem from a diet that includes lots of fiber plant foods like spinach, broccoli, tomatoes, nuts, seeds, etc. Yet meals centered on meat prevail. Researchers are looking at dwindling hunter-gatherers, whose diets traditionally tend to be weighed more to plant foods, to understand how modern influences on these groups affect health (Crittenden & Schnorr 2016). Hunter-gatherers, generally, have a rich microbiome fit for natural plant foods while among North Americans the microbiome is geared toward animal mucus (Fragiadakis, et al. 2019), which can lead to colon cancer and heart disease.

These bits are related, since, as is true:

- We evolved from ape-like humans who ate mostly fruits and leaves.
- Humans have literally eaten some species into extinction.
- Increasing rates of obesity and corpulence are public health concerns.
- Farming animals for flesh and dairy is a prime contributor to climate change.

We cannot continue at this pace without encountering health and climate doom. As implied by Philippa Brakes, et al. (2019), animal cultures, like social learning related to feeding, can impact species survival. This means humans should respect wildlife biodiversity for its own sake as a self-regulating system contributing positively to the forests, oceans, and atmosphere. We can learn from other species that feeding cultures are balanced and healthful, not maladaptive. Humans must tackle obesity, poor health, and climate change from farming and eating animals by grounding their food cultures in plants and conservation. While genotypic and environmental hypotheses account for human food intake, cultural selection also plays an important part. Social pressure is seen in eating habits, modern food practices, and differences in metabolic levels and gut microbiomes across societies leading to obesity (Bajrami 2019).

Cognitive Dissonance

Cousin to confirmation bias is cognitive dissonance. Refer to biologist Robert Trivers' book *The Folly of Fools* (2011) about self-deception. As a species we are deceiving ourselves into obliteration (obesity and climate change) by perpetuating the myth that we are born meat eaters and need to continue consuming excessive quantities of animal flesh and dairy products. This illusory effect works both ways. If you consider yourself progressive and liberal minded but eat meat and dairy, the underlying warrants of this book are directed at you, too. In a tangential line of thought, cognitive and expressive dissonance can also be viewed as an evolutionary stress or fear response from older brain regions. Someone, or a group, shuts down information deemed unacceptable to process since it interrupts the equilibrium of psychobiology (Robinson & Demaree 2007). The larger question I am trying to address is not academic but cultural. We should strive to drift social thinking away from an unhealthy, environmentally degrading, animal cruel meat-based diet to one that is vegan. Since we have evolved other cultural practices and are behaviorally flexible, we can achieve a vegan economy given a collective command of will and modifications in outlook from younger generations.

Cognitive dissonance can be ethical. In negative scenarios about humans using fMRI (functional magnetic resonance imaging), vegans show more empathy than omnivores, say Massimo Filippi, et al. (2010). With negative scenes involving animals, vegans had decreased amygdala activity (threat response) with an increase in frontal lobe activity (emotions and moral judgments). Diets can reflect differing values and beliefs, where the vegans reveal more empathy than omnivores and even vegetarians. Basic social cognition is the recognition of another person or life, and vegans score high on this scale. These findings could go against this book's argument. If omnivores are less sympathetically engaged with environmental or animal care, they are less inclined to change their attitudes, values, or beliefs. I'm hopeful the solution is in the logos appeal of my argument, where health of body (avoiding corpulence leading to cardiovascular disease, etc.) and health of environment (tackling global warming, etc.) will be the main drivers. I don't expect to convince meat eaters on the pathos riddled animal cruelty claim, since the research cited here shows that can be a losing battle.

We cannot make progress if we continually limit our knowledge. The brain functions with neural plasticity, so values and beliefs can be and typically are modified through culture. I am not offering only a biological account for veganism. While evolutionary biology factors into a claim for veganism, it's only the older part of the story. I want to focus on cultural evolution. While human australopith ancestors undoubtedly survived on a diet full of fruits, leaves, nuts, and seeds (Chapter 4), by the time we get to other human predecessors in our bushy pedigree there is a growing reliance on meat. Although vegans are hard pressed to admit some facts, the human gut is not really designed to digest only plants, though it can. Culturally, cooked meat has been a staple food for humans over hundreds of thousands of years. That does not mean, however, that humans are obligate carnivores. Rather, we now need cultural awareness moving away from eating meat and dairy since that diet can engender bad health, environmental degradation, and animal cruelty (Tilman, et al. 2011). Just as multinational corporations and advertisers deceived the public by denying the ill effects of tobacco, opioids, alcohol, and fossil fuels, etc., we continue to face a towering wall of denial about the bad consequences of meat production (Lazarus, et al. 2021), fast-food, and highly processed comestibles. One could draw similar analogies to the manufacturers of firearms and enablers of gun culture.

There's a paradox of thoughtful feeling in most modern, industrial societies. Philosopher Elisa Aaltola (2019) employs the notion of *akrasia* to describe human omnivores. In spite of one's moral knowledge, he or she acts against it. Omnivores know that humans and animals deserve well-being but eat meat and dairy anyway. This behavior goes beyond cognitive dissonance and includes moral reasoning, chiefly relative to climate change. I'd suggest applying a notion of ethical dissonance. The emphasis, Aaltola suggests, is on a loss of empathy for nature. One does not have to be only reasonable but also sentimental. As Aaltola (2015) elsewhere meditates, emotions are strong motivators in moral decision making. If people come to care about their health and the environment in psychologically real ways and feel the responsibility intensely, then perhaps rational claims will be more palatable.

A Culture of Change

Biases and dissonance are part of social reorientation. The processes and dimensions of evolution, whether biogenetic or cultural, suggest that we can adapt. A cultural change away from meat and dairy could be effected through policy, education, and communal action. If one does not like the term vegan, we can coin a new one. I won't lay out any detailed plan in this writing, since my academic area is evolutionary studies, not legislative reform or policy making. Some suggestions have already been made and will be repeated periodically for emphasis. I will also take an analytical look at two pieces of United States legislation, the Green New Deal and the Farm System Reform Act. Whether or not these bills have become law is moot; for our purposes, I use them as illustrations of well-meaning but slightly flawed policy. Crucial here is education of younger people. If we can honestly let young adults know where their food comes from, how it is produced and manufactured, the ingredients, and the environmental impacts of animal farming, then they can decide for themselves what they want to eat.

Alternatives to corporate meat and dairy farming need to be put on the table for children and adolescents. This is where schoolyard gardens and school kitchens to prepare vegan foods could provide effective leverage. We know cultural evolution works in some forms, evident in successful advertising campaigns that dramatically changed perceptions of a product (e.g., Coca-Cola as "The Real Thing") or attitudes about behavior (e.g., the American Express card, "Don't Leave Home Without It"). Let's tackle some of the issues and assumptions behind the claim for an evolutionary cultural shift to veganism.

Policy

A policy issue involves telling someone else what to do. In this case, legislators and educators would be asked to shift their paradigm away from meat and dairy to vegetables and plant-based "meat" and "cheese." At every level, from an individual school to a city council, from a local health organization to a federal agency, concerned leaders would have to introduce a proposal that identifies the health and environmental problems of meat and dairy and offers a feasible solution. Well-funded community gardens employing biocyclic vegan agriculture that excludes any animal fertilizers (Biocyclic Vegan Agriculture 2020) enter the picture here. Assumptions are that under duress a society can change to ameliorate any further or future damage. While there are global problems melded into the effects of meat and dairy production, reform should come from the vast array of local municipalities across the hemispheres. In turn, that might get the federal legislators' attention.

Substantiation

This issue deals with facts, consequences, and the logic of cause-effect. By now most educated and informed people are aware of the dire need to address climate change. These people are also in tune with how animals from every wild or artificial habitat, whether salmon, dolphins, chickens, pigs, sheep, or cows are sentient and sapient. Moving away from animal rights for a moment, I am trying to demonstrate that because our ancestral relatives relied on a mostly plant diet and survived, we can too. It's probably unlikely that meat will be completely eliminated as a food source, but it's a realistic goal to minimize meat and dairy production considering the health risks, environmental damage, and cruelty to animals. One will ask how this is to be accomplished. That's where the policy issues are relevant. Assumptions here might include how the ingenuity and cooperative behavior of humans in communities will enable them to overcome serious health and environmental problems related to public policy. I'm not an advocate of so-called lab meat and would, instead, like a solution that focuses on local expansion and distribution of existing, successful plant foods. For example, if middle or high schools offered a mandatory class, or at least a workshop, in vegan nutrition and had a community garden (indoor or outdoor) and kitchen tied to that, young people would learn lessons of communal, cooperative sustenance.

Evaluation

This issue involves what's good/bad, right/wrong, and effective/ineffective. The purpose here is essentially not to argue for animal rights, as that has been done by thinkers such as Tom Regan (*The Case for Animal Rights*), Bernard Rollin (*Animal Rights and Human Morality*), Christine Korsgaard (*Fellow Creatures*), Barbara King (*Animals' Best Friends*), and Lori Gruen (*Ethics and Animals*), to name only a few. Peter Singer (*Animal Liberation*) does not argue for rights per se since he is a utilitarian. Specifically, Gary Francione (1996) examines the distinctions between welfare (Singer) and rights (Regan). For a good overview of Singer's mostly non-vegan stance and Regan's essentially vegan approach see Josh Milburn (2021), who recognizes the philosophical and political questions of animal rights. Nevertheless, there are socially moral and personally ethical concerns in the claim for a shift to veganism, and those worries are not ignored. If we are desensitized to the emotions, feelings, pain, and suffering of animals, then we've lost what some call humanity. On the moral side, as that word suggests, society needs to shift its focus away from an industry that is harmful to humans, the earth's biosphere, and animals. On the ethical side, each person needs to recognize his or her ability to build a community of vegans for a better future. As for enabling assumptions, readers likely believe that animal lives matter because they think and feel. Animals are not human food, and the destruction of wild animal habitats, along with the development of animal farms, has created an expanding climate catastrophe.

Furthermore, linking substantive, policy, and evaluative issues, zoonotic diseases are engendered from the self-deceptive belief by humans that they need to consume animal flesh in order to survive. One study says the transmission risk of a coronavirus from wildlife to humans is a maximal danger (Nguyen, et al. 2020). Another study shows how pigs are intermediary hosts for a pandemic influenza virus spread to humans (Sun, et al. 2020). Preliminary investigation (Pickering, et al. 2021) reports that SARS-CoV-2 can be spread through pigs. Edward Holmes, et al. (2021) support evidence that the 2020 global pandemic originated in a spillover to humans from live market animals susceptible to a SARS virus, not necessarily from a laboratory leak. A U.N. report (Randolph, et al. 2020) confirms animal to human spread of infectious diseases like Ebola, SARS, and Zika. This last report seeks a proactive effort, which could be supplemented by vegan agriculture in sustainable farming that does not exploit wildlife ecosystems by deforestation. Permanent deforestation across the globe occurs as a result of "commodity production" like cattle farming and other large-scale agriculture (Curtis, et al. 2018).

Sustaining forest biodiversity beneficially nourishes a cleaner global atmosphere, and that attitude is inherent in a vegan ethos.

Evidently, social and cognitive biases affect what people eat, but we can change our minds. With the policy, substantive, and evaluative problems covered so far, let's look at the commercialization of food and its effects on human and environmental health to see how issues tie together. In subsequent chapters, there will be a closer look at the biological, evolutionary, and cultural aspects of food ecology building from generalities in this chapter.

Food Industry

Many might say that leaders and policy makers have not done enough to stem, on the one hand, growing obesity rates and, on the other hand, famine. For this discussion let's focus on what's known. Obesity is on the increase, including all diseases related to corpulence, because of the growing intake of high calorie foods, carbohydrate consumption, starches, and animal products. In the United States alone, from 1999 to 2018, the prevalence of obesity increased among adults, ages twenty and over, up to 40 percent (Hales, et al. 2020). Overeating fats and processed sugars leading to obesity and diabetes is a mismatch from our evolutionary past where those calories were scarce (Griskevicius & Durante 2015). Obesity leads to heart disease, stroke, type 2 diabetes, and some cancers with an estimated annual cost of $142 billion in 2008 U.S. dollars (Gibbons 2013). Contradicting any adaptive fitness, red and processed meats have been determined as carcinogenic links and DNA damaging mutations leading to colorectal cancer (Bouvard, et al. 2015; Gurjao, et al. 2021). The medical sector's response to obesity, diabetes, hypertension, high cholesterol, or other diseases from animal diets or overeating is to prescribe medications. The pharmaceutical industry spends about $240 million each year on at least 1,500 lobbyists who promote drugs over healthy eating (Stone 2011; Wouters 2020). Yet, cancer clinicians are advised to promote a healthy diet that limits or at best excludes red and processed meats (Rock, et al. 2020). Nutritional recommendations by the U.S. Department of Health and Human Services, unsurprisingly, include a variety of diets, one of which is "vegan style" to lower the intake of saturated fats (Gibbons 2013).

To be clear, many diet-related health issues and diseases, such as obesity, were not deliberately chosen by an individual. People don't consciously decide to harm their well-being through food. Most industrialized societies live in a culture dominated by corporate agricultural interests spending much advertising money promoting products causing physical, mental, and social ills. Many urban areas are packed with some inhabitants who don't have access to healthy foods. If we can change the cultural dynamic so that people see there are alternatives to meat and dairy, and if we can get those vegan substitutes into their kitchens through local channels, health outcomes would surely improve. Our food industry is also responsible for pollution and biodiversity degradation. According to a report on climate change and land use (IPCC 2019a), humans directly affect 70 percent of land surface for their agricultural needs, often at the expense of ecosystems that help sustain global climate stability. This report notes that since 1961 vegetable oil and meat production has doubled, increasing unhealthy caloric intake leading to obesity, creating excess foodstuffs that end up as waste, and dramatically adding greenhouse gas emissions reducing air quality. As if this is not enough bad news, the report also notes that over 800 million people are undernourished, in spite of a food glut in some affluent industrialized nations.

The simplest answer is to support and promote a vegan diet free of any animal products, whether flesh or dairy, and allocate excess veggie foods to those in hunger. From a utilitarian perspective, a shift to in-vitro lab meat seems ideal, maximizing the good for the greatest number of people. Wurgaft (2019) wonders if lab meat would change the way we relate to "animals." Probably not, since it's still meat to be eaten, though unnatural in its creation, which could whet appetites for more animal flesh. (The lab meat question will be addressed near the end of Chapter 5.) Rights theory activists might not advocate a lab meat industry since some forms of this process will depend on animal suffering as it breeds them for cells that continually need to be replenished for the laboratory products. Preferable, as practiced by most vegans, is an ethical approach. That is, most vegans are so because they wish to practice virtues of temperance, self-control, and kindness. Not to make a blanket statement, but generally speaking based on the data cited, self-indulgence seems more the norm in many wealthy, industrialized societies. This is part of the bid for the cultural evolution of veganism, where leaders, policy makers, and educators could stress the practical consequences of virtue ethics over a consumerist mentality.

That won't be easy. According to Aristotle, the virtues allied with temperance stem from reason. So the appeal is to people's sense of what is logical: a move away from cruel animal farms and in-vitro lab meat. Both of these practices increase environmental degradation and decrease human health. On average, says *Our World in Data* (2017), a North American will eat about 273 pounds of meat per year, up from about 224 in just a few years (Daniel 2011). As is well known, the body does not store protein from animal flesh, so any excess becomes fat leading to health problems. From a policy position lab meats could be realistic alternatives to omnivores without reducing their liberty to eat meat, but in-vitro meat is not the final solution to climate change, poor health, and cruelty to animals. Good consequences matter, but they should stem from a place of virtuous intent and motive, not selfish consumption at the cost of animal life, human health, or the continuation of tropical forest destruction. In *The Descent of Man*, Charles Darwin (1871, Chapter 4) says that charitable groups are more successful than selfish cheaters. Some readers might be uneasy with talk of virtues, but anyone who is repulsed at the sight of unprovoked cruelty experiences such a virtue. Besides, it's not as if I am talking in abstraction and using a concept like "soul." As the British moralists of the eighteenth century demonstrated, in philosophers like David Hume and Adam Smith, there are innate human tendencies or moral sensations of approval and disapproval. When Australia was burning in 2019, people were distraught at the sight of such immense carnage and devastation where thousands of animals suffered and died, caused by human-induced climate change.

As for taste, whoever says he likes the flavor of meat is referring to the added ingredients, like salt, pepper, and spices on cooked meat. That's more cultural and less biological. No self-proclaimed meat lover would march into an animal farm and gobble up the warm organs of a newly killed pig. The math in this calculation is not complicated. Billions of hungry humans eat billions of dead animals every year, meat pumped with chemicals. Feeding animals for slaughter is a waste of resources. A plate of home-grown roasted, seasoned vegetables would solve this problem. Immense amounts of grain are farmed just to feed these animals, and lakes of water are depleted. In the western U.S., according to one study (Richter, et al. 2020), humans are depleting water faster than it can be replenished, thus damaging ecosystems, and this is due in great measure to raising cattle. The authors of this study conclude that people need to consume less beef to ameliorate the dire water shortage. A similar scenario is playing out in other countries that have deforested lands to raise livestock as human food.

Offal and other pollutants spill into the water tables, other lands, and the oceans. Much of the grain fed to cattle, and perhaps even the water, with the right policy decisions could be distributed to communities in need. In a report on global heating (IPCC 2019b), there is some emphasis on how changes need to be made in agriculture in order to achieve food security, whether in mixed land use, livestock reduction, water efficiency, etc. Sustainability depends on reducing land for meat production in most places (FABLE 2021). Eighteen to 21 percent of water used in food production would be saved in a shift to a vegetarian diet, and there could be a 55 percent lessening of greenhouse gas emissions with a vegan diet, since meat and dairy are a common source of pollution (Al-Delaimy, et al. 2020). More close contact with animals equates to more zoonotic diseases, apparent from the past decade or so (Randolph, et al. 2020). While the cultural evolution shift in attitudes and practices toward veganism could apply to intransigent meat eaters, a prime audience would be young people with conscience and intelligence who desire a less polluted and more stable environment on earth.

No one's nutrition or health will be compromised by the elimination of meat and dairy. Researchers Gidon Eshel and Pamela Martin (2006) say that plant-based diets are "nutritionally superior" to one grounded in meat and dairy. Employment opportunities need not be lost. Just as the fossil fuel jobs should have shifted to wind, water, and solar producing energy under responsible leadership, ranchers and animal farm employers can, with the proper policy guidelines and governmental funding, move to more environmentally friendly vegan agriculture or green energy production. People who live near pig farms tend to be poor, and they suffer under the noxious fumes billowing from these slaughterhouses. That attack on a community's health should matter.

In sum, it's clear that the food industry in most industrialized nations is unhealthy. While I'm suggesting a cultural movement to veganism, leaders could revise planned legislation to include language that points overtly to a vegan culture.

A Look at Some Legislation

Confirmation bias, along with cognitive and ethical dissonance, can appear on the legislative level. What follows in this section is an academic exercise, of sorts, looking at legislation from a critical, historical perspective. Legislation, especially any that purports to address the somewhat ambiguous notion of infrastructure, should include an overhaul that dismantles animal farming and expands plant-based production for the public good. The attention in this section is on policy but includes issues of substantiation.

In an email from October 2019, Senator Charles Schumer (Democrat, N.Y.) says the U.S. Department of Agriculture introduced a new rule eliminating production line speed limits at pig slaughterhouses allowing state inspection officials to be replaced with company employees. There was an effort by members of the House of Congress to block the rule from implementation by introducing an amendment in the next fiscal year's agriculture appropriations. However, this amendment was not included in the final appropriations bill signed into law. Along with high speed slaughter guidelines, concentrated feeding operations have raised animal "welfare" concerns. These operations are large, industrial agricultural facilities that allow a great number of animals, sometimes millions, to be raised at a much faster rate and lower costs. These feedlots were identified as potential pollutants in the 1972 Clean Water Act, since they contaminate groundwater, surface water, and air quality. Schumer asserts that he will keep these issues in mind, mostly during and after the pandemic, to ensure Congress is protecting the heath of Americans, the environment, and all animals.

This approach, along with what follows in the next few paragraphs, might not be good policy from a substantive or evaluative angle.

On the one hand, this email validates what many have told us: animal agriculture cares more about company profits than it does about worker safety, product cleanliness, ecological flourishing, or animal lives. On the other hand, I wonder how much even a few members in the House and Senate of the United States Congress can do. It's a start, but since most people have their attention on social media, television, and their dinner plates, an educated guess is that changes in how we treat health, the environment, and animals must come from a shift in cultural attitudes, and hence my emphasis on transformation emanating from educational institutions or local communities. Social media can have a profound effect on one's beliefs and perceptions of public health (Fuentes & Peterson 2021). A majority of Americans do not monitor the legislative docket in Congress but are focused on happenings in their neighborhood or on their cellphones. Policy makers are part of the solution, but we don't need their permission to act.

The way many people live now can have enormously positive effects for the future in contrast to some feckless, federal legislative battles. That statement is not meant to minimize the sincerity of elected officials, many of whom across continents do have the health and safety of their citizens in mind. However, if by illustration for historical purposes we examine two pieces of legislation, you can see the shortcomings.

While there is much to laud in the Green New Deal (Ocasio-Cortez, et al. 2019), worth supporting, there are some glaring omissions. While the emphasis is on tackling climate change, there is no explicit recognition of how an energy efficient vegan economy could be the driving force in that charge. Instead, there's mention of, for example, a decline in "healthy food." I doubt the drafters of the legislation had vegan agriculture in mind since "healthy" can be an ambiguous term. The bill recognizes how irresponsibly the United States has contributed to global warming, but the call for "healthy food" in sustainable industrial and natural environments falls way short of advocating veganism, local vegetable farms, community and schoolyard gardens, vegan distributors, and vegan food cooperatives, etc. There is an unspecified nod to "community-defined projects," but that could be more explicit and not open to a community starting a chicken farm. Family farming is encouraged, but there's no mention of how large or if vegan, so cattle ranchers and large dairy producers will continue to be subsidized by the federal government over the health of people in a community and the environment. There's passing mention of a "sustainable food system" without naming explicit assumptions. Any such arrangement should be completely plant-based with small outfits in key locales making and distributing lightly processed "meats" like the Field Roast or "cheese" like the Miyoko brands. Any reading of the bill includes the perpetuation of meat and dairy farms even if improperly regulated, evidenced in the Schumer email. That seems to defeat the entire *raison d'être* of the proposed law to address climate change. The bill continually refers to health, which could readily be improved within a generation or less by eliminating processed meat and dairy from the diets of children and teenagers in all schools, not by encouraging the continuation of any form of animal farming.

In the Farm System Reform Act (Booker 2020) there's an attempt to place a moratorium, as outlined in the email by Senator Schumer, on concentrated feeding. Why not cease those operations by using the ingenuity of the modern mind to refabricate meat factories for the production of plant-based products. During World War II and the 2020 pandemic, some factories were reinvented to shift their product outcomes, so the model is there. This bill, rather than promoting veganism, asks for better labeling of meat products. That cataloging action will not necessarily lower the fat and cholesterol intake of citizens. Someone who wants a juicy steak for the grill is not going to read a label. In fact, the bill seems to be much less about animal cruelty and more to ensuring the continuation of animal farming, mainly in the hands of small or family businesses. Any ethnic "mafia" or other large group like a gang could be considered a family. So this is a labor bill, in effect, for cattle ranchers and slaughterhouses that are not part of the mega-corporate machinery and want a piece of the meat pie. Supposedly, this bill, like the New Green Deal, promotes health, but that seems unlikely if it perpetuates meat and dairy production and pollution. If there are ten small, meat or dairy farms in one county, then there's a ten-fold spread in methane gasses, water waste and pollution, and animal offal to deal with, not to mention the carbon emissions for worker and meat transportation.

Livestock pumped with antibiotics and hormones is neither healthy food nor is it the only product a farmer can sell. This unimaginative bill takes aim at corporate agriculture, not health or climate change. Not to be cynical, but it could be the result of lobbyists for small farmers, not vegan advocates.

Whether they've become law or not, the two aforementioned bills simply do not go far enough, and they're only included as examples of ambiguous legislative language. Their warrants should be explicitly vegan. As a matter of public health, legislators in the District of Columbia and New York State are considering laws requiring that doctors receive more training in nutrition and advise their patients about healthy choices for disease prevention, considering high rates of obesity, heart disease, diabetes, and colorectal cancer (Keevican 2019). Regrettably, as with the federal legislation, these bills do not overtly endorse a vegan or even a vegetarian diet.

In the Preface, I call for an awakening about a vegan economy, some form of which we could have in the United States. With veganism, it would be fruitless to go only part of the way when so much is at stake. These bills, in any reading, endorse meat and dairy commerce and do not legislate the kind of revolutionary replacement needed – veganism – which would lead in healing poor health and our damaged environment. As medical doctor Caldwell Esselstyn (Stone 2011) says, most human diseases result from high consumptions of meat, dairy, and processed foods. If a national government wants to prescribe citizen well-being, evident in these two bills, it should favor a vegan culture. With the Farm System Reform Act in particular, any monetary or control redistribution of animal farming does not minimize how many sentient creatures are killed or ultimately consumed. The harmful health and damaging environmental effects are still present, just removed from a few corporate perpetrators and spread more evenly among non-industrial farmers.

In contrast, New Zealand is serious about going green through a reduced reliance on animal agriculture. The problems of livestock farming are many, as that country admits: greenhouse gases, water pollution, deficient sewage systems, soil erosion, toxic fertilizers, deforestation, and biodiversity loss. Wangari Maathai (2006) says these issues should be treated as enemies of a nation. North Americans who eat plant foods contribute significantly less carbon emissions than a meat and dairy eater consuming the same calories (Eshel & Martin 2006). Methane is more dangerous than carbon dioxide when coming from the intestinal emissions of cattle and sheep jammed together. Regarding New Zealand, Jasmijn de Boo and Andrew Knight (2020) talk about "planetary boundaries" since without any, there is no road to sustainable development. Grazing areas could be made wild again to lessen global overheating or converted to eco-friendly places like parks or local farms. A vegan diet uses far less water, land, and crops that are wasted to raise livestock for slaughter. Veganism is not perfect. Nothing is faultless since all living creatures require food substances and then produce wastes, some more injurious than others.

Considering the health and environmental risks of meat and dairy eating, as well as the animal suffering, vegan agriculture would be more advantageous all around and is commercially viable with industry and government support and funding. De Boo and Knight (2020) lay out a "green" blueprint for such an endeavor, beyond the scope of this book and surely more ecologically friendly than the Booker (2020), and to some extent, the Ocasio-Cortez (2019) plans. Let's try to place emphasis on individual vegan agency and responsibility at the community level. As de Boo and Knight say, the most "environmentally-friendly" diet contains no animal products. We can do that on the institutional level from cafeterias in private and public grade schools, museums and universities, to local and federal governments worldwide.

Admittedly, expansive animal agriculture has degraded the environment and biodiversity with an ultimate consequence of disabling plant agriculture itself. Any transformations, say Adam Vanbergen, et al. (2020), are tiered among various solutions: production; attitudes regarding diets and lifestyles; farming scales. These authors claim urbanization is where sustainable vegan values seem to reign, but of course people in many countries don't eat animals for religious or financial reasons. They speculate that improvements in the production of meat alternatives could reduce environmentally detrimental livestock farming, though they don't seem opposed to modest meat consumption. A significant problem is that with 8 billion hungry humans, even a little meat production and consumption would be too much. Newson and Richerson (2021) say there's actually a projected decline in population based on data they cite. This is news to celebrate because that population control is occurring through cultural change. So, too, we can culturally alter destructive dietary habits.

Cultural attitudes of the general public are not so easy to change. Obviously, talk of evolution, our kinship with great apes, and the necessity for a vegan economy pose an existential threat to conservative thinking, as outlined so far. Some researchers (Eadeh & Chang 2020) believe that fear can also be a motivating factor for liberal activism. We should have faith in that observation. Rather than stonewalling, progressive-minded people will seek answers to the dilemmas of bad health and environmental devastation. We can trust they see the solution in a vegan economy. There is no political ownership of issues centered in public health, climate change, or animal cruelty. At least, there should be wide consensus and not partisan politics. One solution is to focus on community involvement about veganism and expand outward from there.

The Moral Complexities of Eating Meat

In an attempt to pull together the various strands of the preceding sections, critical disagreements about the question of morality in eating meat and dairy need to be addressed. The concentration here is more to the evaluative and related substantive issues involved in my argument. Moral or immoral behaviors have consequences worth examining.

Over the past forty years, people have been talking about the ideas of Singer (2009) and Regan (2004), so the cultural climate has shifted marginally. Under natural conditions animals do have legitimate interests in their own lives, those of conspecifics, and their habitats. I'm not making an argument about animal rights and don't want to enter that fray. Rosalind Hursthouse (2000) demonstrates how any moral argument related to animals is open to critique. As philosopher Nathan Nobis (2018) says, questions about using animals in experiments, for fur, or as food often generate the best and worst aspects of anyone's character. As an ethicist, Nobis challenges notions about animal harm. Many people accept that it's wrong to injure animals but participate in the killing because they either don't realize animals are being harmed or think that they can't make a difference. Most meat eaters can be ignorant about what happens on animal farms. The point is that by virtue of cultural evolution the tone and tenor of animal well-being has over a short period of time become part of the debate about health and climate change, as argued here.

Some people defend the production and eating of meat. Others, like Neil Mann (2018), claim that meat has played a "critical role" in the human diet back to 4mya, in spite of, as I will get to in Chapter 4, our mostly frugivorous ancestors. In his own words, Mann's "reality" is a meat-centric hominin. In fact, alarmingly, Mann claims there is no "hard evidence" between meat consumption and coronary heart or cardiovascular disease. On the contrary, see the study by Najjar, et al. (2018). The Academy of Nutrition and Dietetics (Melina, et al. 2016) holds that plant-based diets are not only healthy but also environmentally favorable. Good cardiac function and cerebrovascular health are linked to lifestyles low in meat and high in plant foods, like a Mediterranean diet (Walker, et al. 2021). Brian Wood and Ian Gilby (2017) embrace "man the hunter" and assert we eat more meat than any other anthropoid primate because we evolved for the behavior morphologically, physiologically, and socially. They base their claims more on intermittent chimpanzee and hunter-gatherer behaviors than the fossil record. Chimpanzee "hunting" will be addressed in Chapter 3, but not all chimpanzees, nor hunter-gatherers, are meat crazed, as these authors suggest. They are on point in noting that hunting is a mostly a male dominated activity, but that could be culturally based.

While not advocating intensive factory farming, Christopher Belshaw (2016) nonetheless suggests that "we" are "permitted" to create animals so as to kill and eat them, since we do so with potatoes (his analogy). Granted, he admits meat eating is a moral conundrum. He does not see any ethical problem with the actual eating of meat, just in how we "allegedly" treat animals as food. Belshaw tries to argue, for instance, that meat and dairy farming are necessary to sustain the parasites and insects surrounding livestock, which in turn feed birds. In the wild, with a true ecosystem and not a farming business, all species are interdependent. Belshaw is talking about birds in a humanly fabricated environment that has destroyed their original habitat and then spreads contaminants like fertilizer, insecticide, and tractor exhaust fumes. Curiously, no-till land farming is optimal since it decreases soil erosion and is hence more effective about irrigation, reduces emissions from tilling machinery, and encourages the sequestration of huge amounts of carbon in soils (IPCC 2021).

Another defender of meat, Donald Bruckner (2016), says that vegans wrongly pose vegetables as the only "obligatory" alternative to factory meat. By his accounting, this argument renders vegans as immoral. Bruckner justifies eating meat from hunting, "humanely slaughtered" animals, and roadkill. This thinking neglects the many non-animal meat substitutes available in tofu, tempeh, seitan, bean burgers, mushrooms, etc. To eat roadkill simply proves the point to be made later that our early ancestors scavenged meat from kills by big cats. While Bruckner's is an argument against animal farming, would he argue eating the neighbor's dog (or child) if killed by a drunk driver? Roadkill is not accidental. Developers encroach on wildlife territory and construct roads interfering with their ranging habits. Unwittingly or not, the driver of the car participates in this killing, for argument's sake. Indeed, there are levels of complicity. Opposed to a hunter, the driver does not intentionally kill the innocent and unsuspecting deer.

Some form of meat eating will persist for obligate carnivore pets (cats), zoo animals, or humans on a restricted/special diet. This expediency is only where in-vitro lab meat could enter the picture. Better yet, Bond Pet Foods of Colorado is manufacturing, through biotechnology, an animal-free protein for cats and dogs (Cison 2020). A study (Knight & Satchell 2021) of over 4,000 dog and cat caretakers illustrates that vegan pet foods are acceptably palatable and healthy. Benjamin Wurgaft (2019) says that some people misleadingly talk about a post-animal bio-economy, but for that to happen we'd need a wholly vegan culture, not businesses that feed on animal flesh in whatever form. There's plenty of meat already out there ready to be eaten, according to J. Baird Callicott (2016). In line with Aldo Leopold's (1949) concern for biotic community wholeness over any one individual, where there are species but not individual rights, Callicott argues for eating animals from a mixed community on a family farm. So he, too, in defense of eating meat, somehow manages to justify carnivory by steering clear of advocating in favor of factory farming, where most of the meat consumed comes from. I can't imagine, and don't want to, families in New York City or its suburbs operating backyard farms with animals to be "humanely slaughtered" and eaten. To say "humanely" raised or slaughtered proves the point about how humans lord over other creatures with their supposed sovereign dominion in the distorted ideology of "humanity."

To put it another way, Callicott (2016) suggests that rewilding grazing lands will increase animal life and, therefore, balloon methane gasses in the atmosphere. First, we don't really know that. Second, biological history shows that wildlife contributes to the self-regulation of a biome and does not diminish it, in accordance with the Gaia hypothesis (Lovelock 1979). Third, human production, use, and consumption of meat and dairy are causes of environmental pollution. There's no environmental corruption inherent in the animals or their otherwise thermostatically controlled and uncontaminated ecosystems. Species live in a holistic equilibrium of struggle for survival with each other; humans upset that balance. Rewilding is a realistic process of recovery, restoration, and adaptation but must be based on, among other considerations, biology and ecological sciences, sustainable longevity, collaboration, and ethics (Carver, et al. 2021). Rewilding just 15 percent of the massive lands used in animal farming in key areas would sequester up to 30 percent of CO_2 and prevent up to 60 percent of extinction (Strassburg, et al. 2020). A widely accepted vegan culture would easily enable the rewilding of more than 30 percent of land used in animal agriculture (Theurl, et al. 2020). Passive rewilding works. One way is simply to leave a very large parcel of land alone, and over decades it will regenerate itself into a complex and self-sustaining ecosystem of brush, shrub, or woodland depending on where it's located (Perino, et al. 2019; Broughton, et al. 2021).

Callicott (2016), without seeming to understand that people can choose to avoid eating meat, offers some anecdotes critical of vegans, intimating they are antisocial. Gary Francione (1996) notes how the movement beginning in the 1970s from animal welfare ("humane" treatment) to animal rights ("nonhuman" beings possess inherent value) was perceived as a threat to the heavily funded research and medical communities that labeled animal rights activists as terrorists against humanity. Discrimination against vegans is unjustified second-order intolerance of people opposing animal abuse, a first-order harm (Horta 2018). Repeatedly, Francione goes on, corporate, university, and governmental agency leaders who benefited from cruel experiments on live animals publicly criticized and attacked the views of animal rights activists as radical and contrary to the status quo. Oddly or not, most people would not see a distinction between welfare and rights, debated by scholars. These everyday people know their pets deserve care and have rights, like a family member. Yet these same people eat animals and feed "scraps" of meat to their dogs. Competing frames of mind are cultural and can evolve into meaningful or hurtful practices. What's troubling is how Callicott comes out and says that individual (i.e., ethical for me) choice will not make any difference in the big picture. Multiplied by millions, these "negligible" effects (his word), like gases from each "farmed" cow, would have a cumulative result if reduced year after year. Callicott speaks highly of communitarian ethics, which is exactly what one can foresee in a vegan culture.

Talk about negligible effects is a bit twisted. Since there are so few remaining populations of indigenous people who fish and hunt, on scale they have little negative effect on wildlife if forests could increase in size. Though they are meat eaters, millions of indigenous people with deep cultural histories are "Eating and feeding … without destroying … but maintaining biodiversity" with food systems based in ecology (FAO and Alliance of Biodiversity International and CIAT 2021). Indigenous people employ systems thinking to maintain naturally, not destroy, sustainability (Bardy, et al. 2018). Dwindling tropical forests mean untenable practices like animal domestication and consumption of processed market meats are trending. Regenerating tropical forest lands, decimated for meat production or monoculture farming, could mitigate one-half of anthropogenic greenhouse gases (Goodland & Anhang 2009). One report (Díaz, et al. 2019) is called, in part, "biodiversity and ecosystem services" since the authors show that nature, if well cared for, contributes to human and other species flourishing, evident in how indigenous people typically treat their homelands in a sustainable synergy. In our evolutionary history, we valued nature and permitted resources to replace themselves. What's changed is a corporate, consumerist mentality pushing products over preserving biomes. Sandra Díaz, et al. (2019) support economic incentives not for expanding production activity but for restoring and preserving biodiversity. In line with my thinking, some of those financial inputs should include educational initiatives and public awareness campaigns about a vegan culture.

Readers see the complaint here is with industrialized societies from Asia, Europe, and the Americas, which, in the balance, if converted to veganism can exert an enormous impact to reverse the damage already done. The lingering though robust hunter-gatherers are a "baseline," say Joseph Burger and Trevor Fristoe (2018), revealing how humans can obviously overcome biotic and abiotic limitations. What these authors imply is that hunter-gatherers act on a sustainable, ecological scale whereas people in sprawling urban and suburban areas do not in their reliance on farmed meat and dairy. For example, an organization called Counting Animals claims that a vegetarian diet per person spares at least several hundred animals, from land, water, and sea, each year. A web tool called Vegan Calculator offers a comparable figure. Multiply these yearly numbers per individual by all of the plant eaters combined and we certainly don't have anything negligible regarding environmental reclamation, human health, or animal lives.

To further address the climate impact of eating meat, research shows that even small farms emit unacceptable levels of methane (Wolf, et al. 2017), and grass-fed beef requires larger cattle populations, hence greater pollution (Hayek & Garrett 2018). Environmental damage occurs even in so-called free-range animal farming (Budolfson 2018). Some researchers (Gill, et al. 2010) say that if small livestock farms are to continue, more efficient practices must be followed to mitigate not only local but global climate change. Perhaps it's an unfair analogy, but that's like the misleading argument made by energy producers for mistakenly named "clean" fossil fuels. It's difficult to see on this world map where any individual vegan agency might create a positive, tangible result. The proof, however, is that without a sizable vegan culture so far, the baneful effects are quite obvious. Vasile Stănescu (2010, 2014, 2019) disputes the Edenic myth of local, small, "happy meat" animal farms as sustainable. Further, E. Widiastuti, et al. (2018) demonstrate that air quality in the vicinity of a small dairy farm is relatively poorer than outside its precinct. What all this means is that our food systems and supplies must adapt, and the optimal method is to adopt a vegan culture. There are moral issues involved in the concerns presented here. So-called family farms have become concentrated feeding operations with negative environmental and health consequences.

Individuals, groups, and especially politicians employ defense mechanisms to ignore the ravages of climate change and the dangers of bad health from animal farming. Those who benefit from meat and dairy production have normalized their values over the science of climate and the ethics of how to treat animals. At the same time, science should not be servile to animal agriculture in creating, for instance, an anti-methane gas vaccine for cows. Regarding negligibility, one person might not make a difference, but we know that with the billions of animal products eaten, there are billions of consumers. The motto of one mega fast-food burger chain is something along the line of "billions served." It's not one person. A pervasive vegan culture could certainly alter the balance. Some will say (Budolfson 2016) that a single vegan choice is effectually inefficient, and excuses are then made to eat meat. I'm not talking about a handful of vegans but whole communities that could arise exponentially with the right vision, education, and leadership. To say one person does not make an effective difference leaves out of the equation all those individuals who do buy and eat meat and so finance harms to health, the environment, and animals. Julia Driver (2016) says meat eaters are "complicit" in these injuries. If most of them stop eating meat there is no effectual inefficiency.

To be fair, while vegans seem to make the right ethical choices about diet, they are, probably, like all of us, accountable for secondary (or at least tertiary) actions and decisions that lead to climate change, harm to others, and even animal cruelty. For instance, industries in which investments are made might not be eco-friendly. Vegans make hard choices about where to draw the line in consumer choices. In this book, no judgments are made about meat and dairy eaters, just the hope that some arguments could change their minds. Nevertheless, if a moral individual can act with a like-minded group when connected to similar groups, together they can produce, in the balance, advantageous outcomes for the environment and the health of others, including animals. The expectation is that many vegans can efficiently and effectively create positive change through cultural evolution.

Other points worth considering in this section include the question of whether a meat and dairy eater actually causes animal suffering and death (McPherson 2016). Some people are skeptical of such causation. Bob Fischer (2016) thinks we're permitted to blame meat and dairy eaters, not as individuals, but as a group complicit in immoral structures harming the environment and animals, pointing to the cultural arrangement of food. There certainly is negligence that contributes to the erosion of the health of the individual and the environment. Clayton Littlejohn (2016) calls this type of person, magnified a million times, an "unreflective carnivore." Perhaps this is so because of what Ben Bramble (2016) offers: meat and dairy consumption on a daily basis, contrary to healthy diet guidelines, is an addiction. In fact, some writers (Rowland 2017; Farrimond 2020) say that cheese from animal milk contains casein particles that affect brain cells like an opioid, along with addictive taste kicks. Others might go as far as saying that from birth people addict their children to animal products.

Vegans are not without some blame. Lori Gruen and Robert Jones (2016) suggest a vegan "aspiration" and not a vegan "lifestyle." Note that I object to the word lifestyle and prefer the character-based, morally tinged word ethos. Gruen and Jones are justified to say much vegan thinking is illusory, since forms and variations of animal products appear in nearly everything we use or with which we have contact. In this way, veganism should be seen as a process toward a goal. Like Gruen and Jones, Neil Levy (2016) sees a mistake in vegans promoting their cause as pure; it's not exempt from secondary or tertiary environmental or animal abuses.

For instance, faux meat "burgers" are not necessarily produced, packaged, and disturbed by large corporations in environmentally friendly ways. I hope readers understand that I'm not really pushing a cause or ideology but a proposal to improve human and environmental health, which will result in less human and animal suffering overall. Ethical vegans make a concerted (not perfect) effort to avoid animal products and harms to ecosystems.

According to Alexandra Plakias (2016), it's less an issue of what constitutes food and more about the process a substance undergoes before we call it food. True, but end choices millions of consumers make do have ultimate effects. So-called small organic animal farms are not as humane as they pretend to be. Mobile slaughter units that service small farms rather than having cows transported in bulk still stun the creatures, often several times, and then slit their throats and gut them. While not intensive farming, all of the blood, guts, and offal contribute to the 1.4 billion tons of environmentally dangerous waste produced in this way each year (Gruen & Jones 2016). Uttam Khanal, et al. (2018) ask if smallholder farms adapting some technical improvements addressing climate change boost their efficiency. Fortunately, the answer is yes, but this depends on the farmer's willingness, level of education, as well as market and social capital. If we were to have a cruelty free animal culture, we'd start solving many health and environmental problems, which, in turn, could lead to some positive social change. Questions abound about who'd pay for these mechanical and operational adjustments and why governments are not massively subsidizing vegan agriculture across the industrialized world.

The Question of Ethics and a Social Contract

Obviously some of this discourse is in a moral realm since young people today, even more than before, are exploited by the meat and dairy eating mentality of most corporations and those myopic leaders and policy makers who really ignore climate change. The question is how young people will ethically address the health and climate problems thrust upon them. Philosopher Tyler Doggett (2018) suggests a moral wrong not only in producing and eating meat and dairy but also in the subsequent health and environmental problems generated. He's more concerned with the moral question of meat production, the massive animal suffering, persecution, and slaughter. Additional bad consequences cannot be ignored, like the harms to public health (pathogens) and air and water systems. Some less affluent U. S. Americans in urban areas have no fresh fruit or vegetables, no supermarkets (which are conduits of vegan products), and are forced to eat fatty fast foods. A majority of urban and suburban dwellers across the world have the financial means to make informed dietary decisions. The ethos of the moral individual among a collection of ethical vegans can help move us toward better health, environmental integrity, and less suffering of animals and impoverished people. This might be a type of contractual agreement.

A social contract is justice by mutualistic rules among people for governance outside any natural laws, says Martha Nussbaum (2004). True, but this is artificial and ignores the biology of morality evidenced by altruism, reciprocity, mutualism, and even empathy in the animal world (Tague 2016, Chapter 2). To her credit, Nussbaum is less about a social contract and more to the entitlements of justice for all persons as a human right. Animals and the environment should fit into these privileges. What many people loosely label as "the environment" is a networked collection of individuals, populations, communities, ecosystems, and biomes that constitute the biosphere. If corporations have earned the rights of persons, that privilege should equally apply to ecosystems. If corporations are persons, they should not have the right without penalty to murder animals and kill ecosystems. At any rate, Nussbaum says that where you are born determines your life history, lifespan, education, income, etc. This is true, too, of how the environment and its animal inhabitants are treated. Tropical and rain forests are prime targets for corporate and political looting. Nussbaum is concerned with global inequalities across nations, where self-governance has been "eroded" by big corporations. Instead, she favors a "capabilities approach" that offers basic rights and entitlements to foster human development. The implied question concerns what humans need to live a full life. In Nussbaum's estimation, this approach will balance "fellowship and self-interest" in cooperation. In social contract theory, parties to the agreement are assumed to be equal in powers, but this is not true in a world where wealth and control are concentrated in the hands of a few groups.

Beyond what Nussbaum (2004) suggests, we need to think of ourselves not only as people who want to live peacefully with humans but also as acknowledgers of the sphere of animals in their ecosystems. Similarly, ecofeminism brings to light the moral responsibilities and obligations a society has toward nature, animals, women, children, minorities, and those on the lower rungs of the working class. Feminist animal ethics is opposed to factory farming and animal hunting, gendering these concerns over Singer (the consequences of pain and suffering) or Regan (the rights of animals). Gratuitous violence against women is connected to unjustified attacks on nature, even in the language of exploitation. That is, violence and rational language used to abuse animals is not gender neutral, a thinker like Karen Warren (2000, 2015) might say. Ecofeminism is a multi-vocal practice that makes connections across differing lines of thought concerning the liberation of women, animals, and nature, says Greta Gaard (1993). She follows this line of thinking (2007) regarding environmental feminism and social justice, which, in line with the thrust of my book, could be added to the curricula in many educational institutions. My point is that any social contract should include the rights of ecosystems to thrive as they have for millennia. Human bias and ethical dissonance blind us to the privileges of animal habitats and how those critical environmental spheres effectively help all organisms flourish in ecological harmony.

Furthermore, people on the borders of forests must learn to inhabit shared space with animals. This is a logical claim since "animals" from bacteria and mosses to insects and mammals service the earth's biosphere over land, sea, and air. (Let's exclude deadly viruses.) We can't pretend to have a full life if it involves subjecting ourselves to bad health, spoiling the environment, and persecuting animals. More than any social cooperation, we must shift gears and collaborate with animals, to whom we are embodied and from whom we are nominally separated through evolutionary metamorphoses (Coccia 2021). Therefore, I see "entitlements" for animals since the outcomes that follow will support all life forms. There is no humanity separate from the soil. There's no human "essence" that places us above, say, a mountain gorilla. The shared essence to be valued overall is the energy of life itself, and that's not something a social contract should provoke us to destroy. As Nussbaum offers a plea to policy makers for human rights, why not petition for veganism that ultimately benefits all life.

For example, Maria Voigt, et al. (2018) report that the global demand for natural resources has killed over 100,000 orangutans in Borneo during a period of about fifteen years. To put this in perspective, that's roughly the population of Roanoke, Virginia, or Dearborn, Michigan, or Rialto, California. We could not do without those people who form part of our social fabric. The same can be said for all creatures in an ecosystem. There is no capabilities approach in line with Nussbaum (2004) if orangutans, aptly named people of the forest who maintain foliage sustainably for the benefit of many breathing organisms, are deliberately exterminated.

Much territory has been covered in the opening pages and this introductory chapter to offer background information and a mental template of the health questions, environmental problems, and moral issues. I merely touched on some organic and evolutionary components early on, so now let's zoom into a vegan culture by focusing attention on, in this order, biological factors (Chapter 2), great apes (Chapter 3), early humans (Chapter 4), and cultural theory (Chapter 5) relevant to the cultural ecology of food.

References

Aaltola, Elisa. 2015. "The Rise of Sentimentalism in Animal Philosophy." *Animal Ethics and Philosophy*. Elisa Aaltola and John Hadley, eds. London: Rowman and Littlefield. 201–218.

Aaltola, Elisa. 2019. "The Meat Paradox, Omnivore's Akrasia, and Animal Ethics." *Animals* 9: 1125. doi:10.3390/ani9121125.

Al-Delaimy, Wael K., et al., eds. 2020. *Health of People, Health of Planet and Our Responsibility: Climate Change, Air Pollution and Health*. Cham, Switzerland: Springer.

Alvaro, Carlo. 2020a. *Raw Veganism: The Philosophy of the Human Diet*. London: Routledge.

Asnicar, Francesco, et al. 2021. "Microbiome Corrections with Host Metabolism and Habitual Diet from 1,098 Deeply Phenotyped Individuals." *Nature Medicine*. https://doi.org/10.1038/s41591-020-01183-8.

Bajrami, Ani. 2019. "Cultural Selection and Human Food Preferences." *Journal of Biological Research* 92(7641): 47–49.

Bardy, Roland, et al. 2018. "Combining Indigenous Wisdom and Academic Knowledge to Build Sustainable Future: An Example from Rural Africa." *Journal of African Studies and Development* 10 (2): 8–18. doi:10.5897/JASD2017.0481.

Belshaw, Christopher. 2016. "Meat." *The Moral Complexities of Eating Meat*. Ben Bramble and Bob Fischer, eds. Oxford: OUP. 10–29.

Biocyclic Vegan Agriculture. 2020. www.biocyclic-vegan.org.

Booker, Cory. 2020. *S. 3221.* "To place a moratorium on large concentrated animal feeding operations, to strengthen the Packers and Stockyards Act, 1921, to require country of origin labeling on beef, pork, and dairy products, and for other purposes." https://www.congress.gov/bill/116th-congress/senate-bill/3221/.

Bouvard, Véronique, et al. 2015. "Carcinogenicity of Consumption of Red and Processed Meat." *The Lancet Oncology* 16 (6): 1599–1600. https://doi.org/10.106/S1470-2045(15)00444-1.

Brakes, Philippa, et al. 2019. "Animal Cultures Matter for Conservation." *Science* 363 (6431): 1032–1034. doi:10.1126/science.aaw3557.

Bramble, Ben. 2016. "The Case Against Meat." *The Moral Complexities of Eating Meat.* Ben Bramble and Bob Fischer, eds. 135–150.

Bramble, Ben and Bob Fischer, eds. 2016. *The Moral Complexities of Eating Meat.* Oxford: Oxford UP.

Broughton, Richard K., et al. 2021. "Long-term Woodland Restoration on Lowland Farmland Through Passive Rewilding." *Plos One* 16 (6): e0252466. https://doi.org/10.1371/journal.pone.0252466.

Bruckner, Donald W. 2016. "Strict Vegetarianism is Immoral." *The Moral Complexities of Eating Meat.* Ben Bramble and Bob Fischer, eds. Oxford: OUP. 30–47.

Budolfson, Mark Bryant. 2016. "Is it Wrong to Eat Meat from Factory Farms? If So, Why?" *The Moral Complexities of Eating Meat.* Ben Bramble and Bob Fischer, eds. Oxford: OUP. 80–98.

Budolfson, Mark Bryant. 2018. "Food, the Environment, and Global Justice." *The Oxford Handbook of Food Ethics.* Anne Barnhill, Mark Budolfson, and Tyler Doggett eds. NY: OUP. 67–94.

Burger, Joseph R. and Trevor S. Fristoe. 2018. "Hunter-gatherer Populations Inform Modern Ecology." *PNAS* 115 (6): 1137–1139. doi:10.1073/pnas.1721726115.

Callicott, J. Baird. 2016. "The Environmental Omnivore's Dilemma." *The Moral Complexities of Eating Meat.* Ben Bramble and Bob Fischer, eds. Oxford: OUP. 48–64.

Carver, Steve, et al. 2021. "Guiding Principles for Rewilding." *Conservation Biology* 1–12. doi:10.1111/cobi.13730.

Cison. 2020. "Bond Pet Foods Develops the World's First Animal-free Chicken Protein for Dog and Cat Nutrition." *PR Newswire* 25 August.

Coccia, Emanuele. 2021. *Metamorphoses.* Robin Mackay, trans. Cambridge, UK: Polity Press.

Cook, John, et al. 2016. "Consensus on Consensus: A Synthesis of Consensus Estimates on Human-caused Global Warming." *Environmental Research Letters* 11: 048002. doi:10.1088/1748-9326/11/4/048002.

Crittenden, Alyssa N. and Stephanie L. Schnorr. 2016. "Current Views on Hunter-gatherer Nutrition and the Evolution of the Human Diet." *American Association of Physical Anthropologists* 162: 84–109. doi:10.1002/ajpa.23148.

Curtis, Philip G., et al. 2018. "Classifying Drivers of Global Forest Loss." *Science* 361: 1108–1111.

Daniel, Carrie R, et al. 2011. "Trends in Meat Consumption in the USA." *Public Health Nutrition* 14 (4): 575–583. doi:10.1017/S1368980010002077.

Darwin, Charles. 1871. *The Descent of Man.* London: Penguin Books, 2004.

De Boo, Jasmijn and Andrew Knight. 2020. *The Green Protein Report.* Auckland: Vegan Society.

Declaration of Toulon. 2019. *University of Toulon.* https://www.univ-tln.fr/Declaration-de-Toulon.html.

Díaz, Sandra, et al. 2019. *Summary for Policymakers of the Global Assessment Report on Biodiversity and Ecosystem Services.* IPBES, Bonn, Germany, www.ipbes.net.

Di Paola, Ariana, et al. 2017. "Human Food v. Animal Feed Debate. A Thorough Analysis of Environmental Footprints." *Land Use Policy* 67: 652–659. https://doi.org/10.1016/j.la ndusepol.2017.06.017.

Doggett, Tyler. 2018. "Moral Vegetarianism." *Stanford Encyclopedia of Philosophy*. plato.stanford.edu

Driver, Julia. 2016. "Individual Consumption and Moral Complicity." *The Moral Complexities of Eating Meat*. Ben Bramble and Bob Fischer, eds. Oxford: OUP. 67–79.

Eadeh, Fade R. and Katharine K. Chang. 2020. "Can Threat Increase Support for Liberalism? New Insights into the Relationship Between Threat and Political Attitudes." *Social Psychological and Personality Science* 11 (1): 88–96. https://doi.org/10.1177/1948550618815919.

Eshel, Gidon and Pamela A. Martin. 2006. "Diet, Energy, and Global Warming." *Earth Interactions* 10 (9).

FABLE. 2021. "Environmental and Agricultural Impacts of Dietary Shifts at Global and National Scales." Fable Policy Brief. Paris: Sustainable Development Solutions Network. https://resources.unsdsn.org/environmental-and-agricultural-impacts-of-dietary-shifts-at-global-and-national-scales?_ga=2.100652381.2132615618.1643986960-1694967053.1643 986960.

FAO and Alliance of Biodiversity International and CIAT. 2021. *Indigenous Peoples' Food Systems: Insights on Sustainability and Resilience in the Front Line of Climate Change*. https://doi.org/10.4060/cb5131en.

Farrimond, Stuart. 2020. "Why is Cheese so Addictive?" *Science Focus*, online 8 August.

Filippi, Massimo et al. 2010. "The Brain Functional Networks Associated to Human and Animal Suffering Differ Among Omnivores, Vegetarians and Vegans." *Plos One* 5 (5): e10847. doi:10.1371/journal.pone.00110847.

Fischer, Bob. 2016. "Against Blaming the Blameworthy." *The Moral Complexities of Eating Meat*. Ben Bramble and Bob Fischer, eds. 185–198.

Francione, Gary L. 1996. *Rain Without Thunder: The Ideology of the Animal Rights Movement*. Philadelphia: Temple UP.

Fragiadakis, Gabriela K., et al. 2019. "Links Between Environment, Diet, and the Hunter-gatherer Microbiome." *Gut Microbes* 10 (2): 216–227. https://doi.org/10.1080/19490976.2018.1494103.

Fuentes, Augustin and Jeffery V. Peterson. 2021. "Social Media and Public Perception as Core Aspect of Public Health: The Cautionary Case of @realdonaldtrump and Covid-19." *Plos One* 16 (5): e0251179. doi:10.1371/journal.pone.0251179.

Gaard, Greta, ed. 1993. *Ecofeminism: Women, Animals, and Nature*. Philadelphia: Temple UP.

Gaard, Greta. 2007. *The Nature of Home*. Tucson: U Arizona P.

Gibbons, Gary H. 2013. *Managing Overweight and Obesity in Adults*. U.S. Department of Health and Human Services.

Gill, M., et al. 2010. "Mitigating Climate Change: The Role of Domestic Livestock." *Animal* 4 (3): 323–333. doi:10.1017/S1751731109004662.

Goodland, Robert and Jeff Anhang. 2009. "Livestock and Climate Change." *World Watch*. https://awellfedworld.org/wp-content/uploads/Livestock-Climate-Change-Anhang-Goo dland.pdf.

Greene, Joshua. 2013. *Moral Tribes: Emotion, Reason, and the Gap Between Us and Them*. NY: Penguin Press.

Greenebaum, Jessica. 2012. "Veganism, Identity and the Quest for Authenticity." *Food, Culture and Society* 15 (1): 129–144. doi:10.2752/175174412X13190510222101.

Griskevicius, Vladas and Kristina M. Durante. 2015. "Evolution and Consumer Behavior." *The Cambridge Handbook of Consume Psychology*. Michael I. Norton, Derek D. Rucker, and Cait Lamberton, eds. Cambridge: Cambridge UP. 122–151.

Gruen, Lori, ed. 2011. *Ethics and Animals: An Introduction*. Cambridge: CUP.

Gruen, Lori and Robert C. Jones. 2016. "Veganism as an Aspiration." *The Moral Complexities of Eating Meat*. Ben Bramble and Bob Fischer, eds. 153–171.

Gurjao, Carino, et al. 2021. "Discovery and Features of an Alkylating Signature in Colorectal Cancer." *Cancer Discovery* doi:10.1158/2159-8290.CD-20-1656.

Hales, Craig M., et al. 2020. "Prevalence of Obesity and Severe Obesity Among Adults: United States 2017–2018." *NCHS Data Brief* number 360. Hyattsville, MD: National Center for Health Statistics.

Hart, Donna and Robert W. Sussman. 2005. *Man the Hunted: Primates, Predators, and Human Evolution*. NY: Westview Press.

Hayek, Matthew N. and Rachael D. Garrett. 2018. "Nationwide Shift to Grass-fed Beef Requires Larger Cattle Population." *Environmental Research Letters* 13: 084005. https://doi.org/10.1088/1748-9326/aad401.

Hoffman, Andrew J. 2015. *How Culture Shapes the Climate Change Debate*. Stanford: Stanford UP.

Holmes, Edward C. 2021. "The Origins of SARS-CoV-2: A Critical Review." *Zenodo*. Doi: https://doi.org/10.5281/zenodo.5075888.

Horta, Oscar. 2018. "Discrimination Against Vegans." *Res Publica* 24: 359–373. Doi: https://doi.org/10.1007/s11158-017-9356-3.

Hursthouse, Rosalind. 2000. *Ethics, Humans and Other Animals*. London: Routledge.

IPCC. 2019a. *Climate Change and Land*.

IPCC. 2019b. *Global Warming of 1.5° C*.

IPCC. 2021. *Climate Change 2021: The Physical Science Basis*.

Jost, John T., et al. 2017. "The Politics of Fear: Is There an Ideological Asymmetry in Existential Motivation?" *Social Cognition* 35 (4): 324–353.

Keevican, Michael. 2019. *"Poll: Most Doctors Want to Discuss Nutrition With Patients But Feel Unprepared."* Physicians Committee for Responsible Medicine, online 13 November.

Khanal, Uttam, et al. 2018. "Do Climate Change Adaptation Practices Improve Technical Efficiency of Smallholder Farmers? Evidence from Nepal." *Climate Change* 147: 507–521. https://doi.org/10.1007/s10584-018-2168-4.

Kim, Hyunju, et al. 2019. "Plant-based Diets Are Associated With a Lower Risk of Incident Cardiovascular Disease, Cardiovascular Disease Mortality, and All-cause Mortality in a General Population of Middle-aged Adults." *Journal of the American Heart Association* 8: e012865. doi:10.1161/JAHA.119.012865.

King, Barbara J. 2021. *Animals' Best Friends: Putting Compassion to Work for Animals in Captivity and in the Wild*. Chicago: U Chicago P.

Knight, Andrew and Liam Satchell. 2021. "Vegan Versus Meat-based Pet Foods: Owner-reported Palatability Behaviours and Implications for Canine and Feline Welfare." *Plos One* 16 (6): e0253292. doi:10.1371/journal.pone.0253292.

Korsgaard, Christine M. 2018. *Fellow Creatures: Our Obligations to Other Animals*. Oxford: OUP.

Lazarus, Oliver, et al. 2021. "The Climate Responsibilities of Industrial Meat and Dairy Producers." *Climate Change* 165: 30. https://doi.org/10.1007/s10584-021-03047-7.

Leopold, Aldo. 1949. *A Sand County Almanac*. Oxford: OUP.

Levy, Neil. 2016. "Vegetarianism: Toward Ideological Impurity." *The Moral Complexities of Eating Meat*. Ben Bramble and Bob Fischer, eds. 172–184.

Littlejohn, Clayton. 2016. "Potency and Permissibility." *The Moral Complexities of Eating Meat*. Ben Bramble and Bob Fischer, eds. 99–117.

Lovelock, James. 1979. *Gaia: A New Look at Life on Earth*. Oxford: OUP, 2000.

Maathai, Wangari. 2006. *The Green Belt Movement*. New Revised Edition. NY. Lantern Books.

Mann, Neil J. 2018. "A Brief History of Meat in the Human Diet and Current Health Implications." *Meat Science* 144: 169–179. https://doi.org/10.1016/j.meatsci.2018.06.008.

Martens, Pim. 2020. *Sustanimalism: A Sustainable Perspective on the Relationships Between Human and Non-human Animals.* De Biezen, Netherlands: Global Academic Press.

McCright, Aaron M. and Riley E. Dunlap. 2011. "Cool Dudes: The Denial of Climate Change Among Conservative While Males in the United States." *Global Environmental Change* 21 (4): 1163–1172. doi:10.1016/j.gloenvcha.2011.06.003.

McDonald, Barbara. 2000. "'Once You Know Something, You Can't Not Know It': An Empirical Look at Becoming Vegan." *Society and Animals* 8: 1–23.

McPherson, Tristram. 2016. "A Moorean Defense of the Omnivore." *The Moral Complexities of Eating Meat.* Ben Bramble and Bob Fischer, eds. 118–134.

Mehta, Tushar and Nicholas Carter. 2021. *Plant Based Data.* https://www.plantbaseddata.org/.

Melina, Vesanto, et al. 2016. "Position Paper." *Journal of the Academy of Nutrition and Dietetics* 116(12): 1970–1980. https://doi.org/10.1016/j.jand.2016.09.025.

Milburn, Josh. 2021. "The Analytic Philosophers: Peter Singer's *Animal Liberation* and Tom Regan's *The Case for Animal Rights.*" *The Routledge Handbook of Vegan Studies.* Laura Wright, ed. Abingdon: Routledge. 39–49.

Miller, Jon D., et al. 2021. "Public Acceptance of Evolution in the United States, 1985–2020." *Public Understanding of Science.* doi:10.1177/09636625211035919.

Najjar, Rami S., et al. 2018. "A Defined, Plant-based Diet Utilized in an Outpatient Cardiovascular Clinic Effectively Treats Hypercholesterolemia and hypertension and Reduces Medications." *Clinical Cardiology* 41 (3): 307–313. doi:10.1002/clc.22863.

Newson, Lesley and Peter Richerson. 2021. *A Story of Us: A New Look at Human Evolution.* Oxford: OUP.

Nguyen, Quynh Huong, et al. 2020. "Coronavirus Testing Indicates Transmission Risk Increasing Along Wildlife Supply Chains for Human Consumption in Viet Nam 2013–2014." *bioRxiv.* Doi: https://doi.org/10.1101/2020.06.05.098590.

Nobis, Nathan. 2018. *Animals and Ethics.* Atlanta, GA: Open Philosophy Press.

Nussbaum, Martha C. 2004. "Beyond the Social Contract: Capabilities and Global Justice." *Oxford Development Studies* 32 (1): 3–18.

Ocasio-Cortez, Alexandria. 2019. *H. Res. 109.* "Recognizing the Duty of the Federal Government to Create a New Green Deal." https://www.congress.gov/bill/116th-congress/house-resolution/109.

Oreskes, Naomi. 2018. "The Scientific Consensus on Climate Change: How Do We Know We're Not Wrong?" *Climate Modelling.* Elisabeth A. Lloyd and Eric Winsberg, eds. Cham Switzerland: Palgrave. 31–64.

Our World in Data. 2017. https://ourworldindata.org/grapher/meat-supply-per-person.

Pennycook, Gordon, et al. 2019. "On the Belief That Beliefs Should Change According to Evidence." *PsyArXiv.* https://doi.org/10.31234/osf.io/a7k96.

Perino, Andrea, et al. 2019. "Rewilding Complex Ecosystems." *Science* 364 (6438): eaav5570. doi:10.1126/science.aav5570.

Pickering, B.S., et al. 2021. "Susceptibility of Domestic Swine to Experimental Infection with Severe Acute Respiratory Syndrome Coronavirus 2." *Emerging Infectious Diseases* 27 (1): 104–112. https://dx.doi.org/10.3201/eid2701.203399.

Plakias, Alexandra. 2016. "Beetles, Bicycles, and Breath Mints: How 'Omni' Should Omnivores Be?" *The Moral Complexities of Eating Meat.* Ben Bramble and Bob Fischer, eds. 199–214.

Potts, Annie and Philip Armstrong. 2018. "Vegan." *Critical Terms for Animal Studies.* Lori Gruen, ed. Chicago: U Chicago P. 395–409.

Randolph, Delia Grace, et al. 2020. *Preventing the Next Pandemic: Zoonotic Diseases and How to Break the Chain of Transmission.* Nairobi, Kenya: United Nations Environment Programme.

Regan, Tom. 2004. *The Case for Animal Rights.* Updated edition. Berkeley, CA: U California P.

Richter, Brian D., et al. 2020. "Water Scarcity and Fish Imperilment Driven by Beef Production." *Nature Sustainability* 3: 319–328. https://doi.org/10.1038/s41893-020-0483-z.

Robinson, Jennifer L. and Heath A. Demaree. 2007. "Physiological and Cognitive Effects of Expressive Dissonance." *Brain and Cognition* 63 (1): 70–78. doi:10.1016/jbandc.2006.08.003.

Rock, Cheryl L., et al. 2020. "American Cancer Society Guidelines for Diet and Physical Activity for Cancer Prevention." *CA: A Cancer Journal for Clinicians* 70: 245–271. doi:10.3322/caac.21591.

Rollin, Bernard E. 2006. *Animal Rights and Human Morality.* Third edition. Amherst, NY: Prometheus Books.

Rowland, Michael Pollman. 2017. "This is Your Brain on Cheese." *Forbes*, online 26 June.

Safina, Carl. 2015. *Beyond Words: What Animals Think and Feel.* NY: Picador.

Schumer, Charles E. 2020. *Email to Gregory F. Tague.* 20 May.

Singer, Peter. 2009. *Animal Liberation.* Updated 1975 edition. NY: Ecco.

Stănescu, Vasile. 2010. "'Green' Eggs and Ham? The Myth of Sustainable Meat and the Danger of the Local." *Journal for Critical Animal Studies* 8 (1/2): 9–32.

Stănescu, Vasile. 2014. "Crocodile Tears, Compassionate Carnivores, and the Marketing of 'Happy Meat.'" *Critical Animal Studies: Thinking the Unthinkable.* J. Sorensen, ed. Canadian Scholars P. 216–233.

Stănescu, Vasile. 2019. "Selling Eden: Environmentalism, Local Meat, and the Post Commodity Fetish." *American Behavioral Scientist* 63 (8): 1120–1136. doi:10.1177/0002764219830462.

Stanovich, Keith E. and Richard F. West. 2007. "Natural Myside Bias is Independent of Cognitive Ability." *Thinking and Reasoning* 13 (3): 225–247.

Stone, Gene, ed. 2011. *Forks Over Knives: The Plant-based Way to Health.* NY: The Experiment.

Strassburg, Bernardo B.N., et al. 2020. "Global Priority Areas for Ecosystem Restoration." *Nature* 586: 724–729. https://doi.org/10.1038/s41586-020-2784-9.

Sun, Honglei, et al. 2020. "Prevalent Eurasian Avian-like H1N1 Swine Influenza Virus with 2009 Pandemic Genes Facilitating Human Infection." *PNAS.* https://doi.org/10.1073/pnas.1921186117.

Tague, Gregory F. 2016. *Evolution and Human Culture.* Leiden: Brill.

Theurl, Michaela C., et al. 2020. "Food Systems in a Zero-deforestation World: Dietary Change is More Important Than Intensification for Climate Targets in 2050." *Science of the Total Environment* 735: 139353. doi:10.1016/j.scitotenv.2020.139353.

Tilman, David, et al. 2011. "Global Food Demand and the Sustainable Intensification of Agriculture." *PNAS* 108 (50): 20260–20264.

Trivers, Robert. 2011. *The Folly of Fools: The Logic of Deceit and Self-Deception in Human Life.* NY: Basic Books.

Vanbergen, Adam J., et al. 2020. "Transformation of Agricultural Landscapes in the Anthropocene: Nature's Contributions to People, Agriculture and Food Security." *Advances in Ecological Research.* David A. Bohan and Adam J. Vanbergen, eds. Academic Press. 193–253.

Voigt, Maria, et al. 2018. "Global Demand for Natural Resources Eliminated More Than 100,000 Bornean Orangutans." *Current Biology* 28: 761–769. https://doi.org/10.1016/j.cub.2018.01.053.

Walker, Maura E., et al. 2021. "Associations of the Mediterranean-Dietary Approaches to Stop Hypertension Intervention for Neurodegenerative Delay Diet with Cardiac Remodelling in the Community: the Framingham Heart Study." *British Journal of Nutrition* 1–9. doi:10.1017/S0007114521000660.

Warren, Karen. 2000. *Ecofeminist Philosophy*. Lanham, MD: Rowman and Littlefield.

Warren, Karen. 2015. "Feminist Environmental Philosophy." *Stanford Encyclopedia of Philosophy*. Plato.stanford.edu

Widiastuti, E., et al. 2018. "Impact of Small Holder Dairy Farm on the Air Quality in Gunungpati District, Semarang Municipality." *IOP Conference Series: Earth and Environmental Science* 119: 012064. doi:10.1088/1755-1315/119/1/012064.

Wolf, Julie, et al. 2017. "Revised Methane Emission Factors and Spatially Distributed Annual Carbon Fluxes for Global Livestock." *Carbon Balance and Management* 12 (16). doi:10.1186/s13021-017-0084-y.

Wood, Brian M. and Ian C. Gilby. 2017. "From *Pan* to Man the Hunter: Hunting and Meat Sharing by Chimpanzees, Humans, and Our Common Ancestor." *Chimpanzees and Human Evolution*. Martin N. Muller, et al., eds. Cambridge, MA: Harvard UP. 339–382.

Wouters, Olivier J. 2020. "Lobbying Expenditures and Campaign Contributions by the Pharmaceutical and Health Product Industry in the United States, 1999–2018." *JAMA Internal Medicine* 180 (5): 688–697. doi:10.1001/jamainternmed.2020.0146.

Wurgaft, Benjamin Aldes. 2019. *Meat Planet: Artificial Flesh and the Future of Food*. Oakland, CA: U California P.

2

BIOLOGICAL THEORY

Since we are products of biological evolution and require food to survive, basic principles of biology and physiology must be addressed, first, in this chapter, followed by consideration of great apes and our hominin ancestors, in the next chapters, before striking into the final chapter on cultural evolution. As has already been intimated, cultural evolution implies biology, and so why the dynamics of science factor into this debate about vegan culture.

Genes respond to environment, and if the mutation is beneficial, it can be passed on and shared in a population as an advantageous adaptation. The evolution of forms and behaviors is not only biological; beneficial alterations can occur culturally. Those simple sentences are at the empirical foundation of this book and expounded upon in the next paragraph.

One could paraphrase a number of people by saying that scientific theories make predictions to exclude the improbable. Science is about understanding patterns. In this vein, evolutionary theory is built on three pillars following Charles Darwin (1859): variation (or mutation), competition (or selection), and inheritance (or retention). Biological evolution comes through the natural selection of the fitter, an individual with an enhanced ability to survive and reproduce, by means of adaptation. Evolution is a change in gene frequencies, or as Darwin says, descent with modification. Invoking concepts from philosopher of biology David Hull, sociologist Marion Blute (2010) relates how evolution consists of variable replicators (genes), selected interactors (organisms), and evolving lineages (species). Per evolutionary psychologists Leda Cosmides and John Tooby (1992), natural selection is a shaping process correlating an organism's form and function. While environments exert a powerful force on organisms (Clark, et al. 2018; Fisher 2019), selection is a regulating factor. For Darwin, adaptation follows from natural selection enabling an organism to fit into its environment; alternatively, an adaptation can be stabilization against mutation (Niklas 2016).

DOI: 10.4324/9781003289814-3

Descendant traits are modified through an ancestral genome accentuated by chance or from necessity. An essential consideration for this discussion is in how the struggle for survival through variation impacts descendants. Approximately 90 percent of all life forms have gone extinct, suggesting that adaptations are temporary and not permanent solutions. The evolutionary case for veganism investigates not only how humans will adapt to their predominantly failing food ecology and worsening climate change, but also looks at the impacts artificially and naturally selected adaptations have on their progeny and organisms in ecosystems upon which humans depend.

With the theory of natural selection, Darwin (1859) explains the inevitable mechanism by which species change over time in order to adapt when there is variation of individuals in a population. Descent with modification means differences have a genetic base so they can be passed on. Heritable diversity in a population is not only genes but can be, as related in this book, values, beliefs, and practices. Biologists (Bartee, et al. 2017; Clark, et al. 2018) tend to describe an organism's beneficial mutations for adaptation as enabling the individual or group – a population's genetic change over time – to fit better into its environs. Ecological conditions, however, can change, which could lead to further adaptations for fitness enhancement and survival or not. Ideas like these will be considered, as well as, in a more muted way, multilevel selection, which benefits a group regardless of effects on individuals (Fisher 2019). Elements of sociocultural evolution, with foundations in Darwinian natural selection and descent with modification, also play a part in this overall discussion but expressly in Chapter 5.

Measurements of Evolution

Evolution can be measured in various ways. Here, a brief overview will be provided, mostly of the biological aspects, and should be kept in mind while reading Chapters 3 and 4. Later, in Chapter 5, cultural evolution will be covered. Biologists Eva Jablonka and Marion Lamb (2005) suggest four dimensions of evolution: genetic, epigenetic, behavioral, and symbolic. Let's touch on these areas, but readers might ask why they are important for this discussion. Human genes are encoded with information that keeps us alive and directs our actions. While we have no control over respiration or digestion, which seem determined by genetic functions, we can make choices about the quantity and quality of what we eat, given what's available. Technically, our genes, not us, control our behaviors. I treat this in more detail in *Making Mind* (2014), but suffice it to say here that one is born with a temperament influenced by external factors. No person is born exclusively as a vegan or meat eater, but as organisms, people need nutritious food to survive. There is a biology that is human much as there is biology for any other species. We are genetically close to great apes, but clearly there are telling differences, though I am not favoring one species over another.

The environment to which genes respond is a fluid context of family psychology, physical and social surroundings, and the changing behaviors of others.

Internal genes function in reactions to external dynamics. Those are only a few qualifiers under the broad umbrella term of environment. No single gene is responsible for any function or trait. Genes interact. At the same time, this gene shuffling or variation that occurs in the creation of each being accounts for individual differences. There are codes, say Jablonka and Lamb (2005), in DNA that can switch genes on/off (Bartee, et al. 2017), and this toggling mechanism might account for the close similarities and yet differences between humans and great apes. A key point made by Jablonka and Lamb is that there is no necessary, causal relation between a gene sequence (biological construction) and traits (behavior). Rather, a pattern or number of genes plus the environment create a brew of potential behaviors. So we are genotypes (biological evolution), but we can change our behavior (cultural phenotypes). This is my focus, too. Biology is not necessarily Calvinist. We can change what we grow as food and what we decide to eat. We are omnivores who can and should become vegans for health and ethical reasons relating to the environment and animals.

Biologists, for example, talk about developmental plasticity, which is the individual's genetic inheritance along with response to external pressures. As Jablonka and Lamb (2005) say, identical genes can produce different phenotypes (individuals) and unlike genes can yield a similar phenotype. Cell dynamics and regulation, they note, account for phenotypic traits and behavior, certainly not DNA alone. Rather than a machine with fixed gears, complex organisms have amoeba-like transcription factors that develop (Bartee, et al. 2017). In wealthy, industrialized societies with so many food options, humans can choose and change behaviors they seem to believe are fixed, like what to eat. Jablonka and Lamb don't necessarily see mutations as totally random. There might be some regulation, as in by analogy, sexual reproduction for the survival of a species opposed to asexual reproduction that can weaken a gene pool into extinction. Extending that analogy further, I'm suggesting that our genes can culturally recombine with less damaging mutations, behaviorally and physically, with a decision to go vegan. Any fixed outlook is unprepared for dealing with unexpected stressors. If untested, creative solutions can be risky. We know how in our primate heritage we evolved from a plant-based diet (Chapter 4). The best course of action in stress, say Jablonka and Lamb, is one of *interpretation*. We must use old wisdom and new ideas, a response to biological strategies for adaptation. Interpretation like this can be intuitive as much as intellectual. Based on what you've read so far, you see what's implied. Our future fitness hinges on evolving a vegan culture to which, as omnivores, we are preadapted.

Another form of evolution involves epigenetics where the transmission and inheritance of non-DNA information comes through cells (Bartee, et al. 2017; Clark, et al. 2018). A famous example of this was the Dutch famine of World War II during the winter of 1944–1945. Prenatal babies suffered health effects from their mothers' starvation, and those adverse growth signatures were subsequently transferred to the next generation. In this case, more at the phenotype than the genotype, variations open to function can happen quickly in response to environmental stress. Darwinian evolution is possible via generational behavior or even learning. Culture can be independent of genes, but the two interact. The acquisition, eating, and sharing of food can influence behavior. Transmissible variations can result in not only a cultural but also a biological change. Part of culture is learning, imitation, and reconstruction, and we need to alter our behavior in such ways regarding the conceptual necessity (or not) and presentation of meat and dairy as essential food products. Protein and nutrients come from a variety of sources, as the diets of great apes attest. The Physicians Committee for Responsible Medicine (2020) flatly claims that meat is not essential for the human diet.

As omnivores, our diets are modular, not fixed, so there are variant behaviors. Variation can affect more than one behavior. While any story of a hunter might focus on the animal kill, some listeners might hear how resources are shared to benefit a large group. In other words, as suggested here, alternative niches can be constructed in which to live and from which to eat. There will be more to say about niche construction later. Currently, many people will not relinquish their choice cut steaks, but some will, and the latter's pattern of behavior can, perhaps, spread quickly into a new generation of meatless eaters. Through social conditioning, a meat eater typically convinces herself that she requires a cowboy or tomahawk steak as essential food and is unlikely to alter her behavior. Others might be open to change, at first reducing meat eating, then eliminating red meat, then going vegetarian, and then becoming vegan. This pattern, if copied by others on a large scale, could have massively beneficial impacts on human health and environmental sustainability. Education and public awareness using information as well as the arts are components of achieving this important cultural variation. Those delivering the messages need not be strict vegans.

In many ways, for the Porterhouse or T-bone steak eaters, the information transmitted in this argument is invisible. For others, the evidence is latent and can manifest itself. Symbolism can be, according to Jablonka and Lamb (2005), a powerful aspect of cultural evolution. As is true in biological evolution, the environment, broadly conceived, is a force of selection, and this is no less true cross-culturally today. Which variants will survive: eating meat and dairy that degrade soil, air, and water or vegan diets that are sustainable? The answer might depend on how these images are represented to young people. As in a large community of altruists and cheaters, while some cheaters will statistically survive, the altruists will ultimately prevail because it is a better evolutionary strategy. So, too, speculatively, meat and dairy eating habits will soon become minimized by many people as environmentally unsustainable. That negative image is gaining momentum. Some will persist, harming their own health, but the tide is turning, pronounced especially by the 2020 coronavirus, and similar pathogens previously, that arose from one of many so-called wet markets oozing animal blood and feces ripe with disease where live, small animals are messily decapitated with a cleaver one after another with little regard for sanitation. In spite of the pandemic, many of these markets, as far as we can tell, persist in their dirty business. Food is not just culture; food is ecology, both powerful and potentially competing symbols.

Jablonka and Lamb (2005) say variation is not random but targeted to help organisms adapt and survive better (Bartee, et al. 2017). The question, then, is how humans will filter their current eating habits and revise them to veganism. Epigenetics and behavioral systems are not outcomes of evolution but the movers of adaptive change. Kate Distin (2011, Chapter 3), for instance, reports on research that shows how ecological, behavioral, and cultural inheritance systems affect human thinking and agency. Low cortisol levels were detected in babies born of Holocaust and 9/11 survivors. Epigenetic ethical rules that affect our behaviors allow us to learn what to avoid (e.g., incest) and how to act altruistically (Ruse & Wilson 1985). A key question is how younger generations will modify their environment of adaptation for the benefit of themselves and their progeny in the future. Since altruism provides fitness enhancing advantages, a cultural shift can advance us away from the self-centered harms of meat and dairy toward the social and biological benefits of a vegan economy. While natural selection has helped evolve our socially moral behaviors, as humans we decide right/wrong, not natural selection per se.

Evolution and Natural Selection: Individuals and Groups

Since some basic concepts of biological evolution have not changed, I'll use some classic texts as the foundation of talking points in the next few sections, including references to more recent research. Mine is an argument for cultural evolution where I use biology analogously.

In one model text, *The Origins of Adaptations*, Verne Grant (1963) relates how animals are not fixed machines but cells and chemicals integrated holistically into their environment. This is also largely covered in Clarence Lehman, et al. (2019) and Matthew Fisher (2019). Similarly, Karl Niklas (2016) discusses plant evolution. Natural selection is not creative; it's a sifting process of mutations and variations that can affect the gene frequencies in a population (Bartee, et al. 2017; Clark, et al. 2018). How any individual survives well to be able to contribute fitness to the next gene pool is vital. Every organism seeks control in its ecological niche and maintains homeostatic equilibrium. Modern humans, in unchecked population growth and industrialization, ignore this biological principle. There was an oil crisis in the 1970s, and rather than shift to alternative energy sources then, like solar, wind, and water power, oil and gas drilling resumed to fuel yet bigger cars and fatter industry. Deforestation to raise palm oil or beef and dairy cattle also exploded during this period, even though there were cries to save the Indonesian and Amazonian rain forests. The entire surface of the earth is inhabited by living organisms ranging from lichens to whales, and each has evolved special adaptations to thrive successfully in its tiny or large habitat. This organic balance was true of early humans, North American indigenous people, and today of remaining hunter-gatherers, but has since been annihilated, mostly within the past few hundred years. The genotype permits the phenotype to adjust responses from a norm. Subsequently, genotypes can be modified by the environment into phenotypic variations, which can be adaptive. This is really true in plants, notes Grant (1963) and Niklas (2016).

Through epigenetic and symbolic/cultural adaptation the human genotype could steer away from self-deceptive meat dependency and morph into a greener phenotype. In a similar way, Gary Hatfield (2013) says niche construction occurs when some organisms alter parts of the environment to establish surroundings that are more amenable. In turn, this non-genetic material becomes ecological inheritance. Selective retention then defines the population. For most modern human societies the niche is fabricated into meat and dairy consumption, and this habitat is inherited and passed along if not rejected. One could say the meat and dairy eating habit is a combination of social facilitation and emulation. If humans can ratchet up learning, it stands to reason that people could learn how to achieve better health, help stem climate change, and eliminate animal cruelty by creating a vegan niche.

An organism selects and is selected by its environment in terms of tolerance and adjustment (Clark, et al. 2018). For humans, these natural boundaries have been so eradicated that we've put in danger many plant species (e.g., tropical and rain forests) and have driven or hunted many animals into extinction. Genes, not traits, get passed on, but the environment as broadly defined, physical, psychological, and social can act so influentially, almost like genetic material, to produce behavioral change. Humans have caused climate damage, a negatively disruptive factor on all land, air, and sea life. A large percentage of this corrosion comes in the form of farming meat and dairy. Stable food communities survive better and longer than unstable (Lehman, et al. 2019; Fisher 2019). A way back to stability is by farming and distributing locally and by eating more plant foods manufactured on a scale that pollutes far less. We presently require a behavioral adaptation.

In another classic text, George Williams (1966) says the process of adaptation can create costs up front and so must arise from necessity if not by chance. See, too, Lisa Bartee, et al. (2017) and Mary Ann Clark, et al. (2018) who discuss adaptive processes. In a parallel manner, Karl Niklas (2016) homes in on plant morphology and physiology. That cautionary burden seems to apply in the line of thinking outlined in this book. Rather than relying on some random events in nature to save earth, we must evolve a course of physical, psychological, and social adaptations with a widespread move to veganism. As Williams suggests, the fitness of a group can increase through adaptations of individuals. With more vegans, advantages fall to groups, so it's reasonable to say we need many ethical vegans, who tend to be environmentally caring and energy efficient, clustered across populations in many global areas. Granted, natural selection is not future-looking. Just as there is no determinism in evolution, there is no providence. As readers see by now, we're speaking of evolution in multiple manifestations, where natural selection is but one.

Natural selection is not inevitable, but climate catastrophe is, unless, like genes, peoples' attitudes and habits segregate and recombine in a more holistically adaptive manner. As natural selection deals with alternatives for fitness (survival and reproduction), humans need to embrace a cultural form of selection in order to ensure their longevity on earth. Selection itself does not see extinction, but we can. Selection is not progressive but a tendency toward specialization as needed for adaptation (Hodge & Radick 2003; Clark, et al. 2018). We need to follow that lead, culturally. Williams (1966) and to some degree Niklas (2016) understand natural selection less as an agent of change and more as the maintainer of adaptations. While this is truthful, consider the alterations that can be achieved in one generation artificially. At this point in human descent our modification of diet might be artificial, but it is ultimately a selection process for overall survival.

Without oversimplifying Williams (1966), there's an evolutionary process and an evolutionary outcome. For the purposes presented in this text, let's say processes could include individual variation, selection among groups, and diversity among populations. These could have outcomes affecting organic adaptation and biotic evolution. Organic adaptation is ultimately on the individual level, but it can accommodate inclusive fitness for a group. Dimensions of evolution confer changes to the flora and fauna of an entire area. At the rate we're going, there seems to be a negative reversal of organic adaptations exclusively for individual groups at the expense of biotic evolution. Stated directly, many humans are so self-interested in consuming all types of animals, animal materials, or earth's resources whether to eat or use with minimal conservation considerations, that some biotas are threatened to extinction. The irony, of course, is that with a rapid-fire elimination of biodiversity, humans themselves are in peril, already evident from the immediate effects of climate change.

According to authors like Williams (1966) and Niklas (2016), the key in natural selection is not merely to survive but to evolve adaptations for genetic longevity through inheritance. See, too, Lisa Bartee, et al. (2017) and Mary Ann Clark, et al. (2018) for coverage. In sexual selection theory, males will enhance their fitness, or appear to do so, in order to gain the notice of discerning females. Apply this by analogy to the argument presented here. By overeating meat and dairy billions of times per year, many well-fed humans are deceiving themselves and others into reproductive (i.e., sustainable) success. Conspicuous consumption of expensive animal foods might appear attractive in the short term, but it's baneful in the long run. There is no future health for human life in a barren world, a planet devoid of resources, whether by depletion or environmental degradation. Like Williams, I intuitively lean to individual selection. Even among so-called pack animals, the competition for resources and mates is on an individual level. That means the group or community is a consequence of individuals mostly acting with self-interest among others.

This apparently cynical evaluation does not exclude altruism, empathy, sympathy, sharing, or donors. Certainly, much animal social life springs from the mother–infant bond. Groups are collections of individuals acting and reacting like one organism, so the ultimate question regards how splinter groups will cohere to form larger, beneficial organizations. The point is that in spite of praise for human cooperation, there is less collective action than we truly need. Governments and more corporations could act mutually in efforts for public health and conservation.

That idealism rubs against the grain of profiteering, the basis of most corporate and some political life. The net result of self-interest, on the individual, group, community, state, or corporate level could be a decrease in the quality of life globally.

Animal groups are self-sufficient. Human nations appear cooperative, but yet vie to out-compete each other in the rapacious pillage of natural resources. In his *Beagle* voyage, Darwin (1839, Chapter 17) affirms how humans willfully and wantonly plundered species from birds to tortoises for food or sport. When the president of any country blurts out, "Nation First," promoting offshore drilling or the mining of minerals from protected lands, that agenda does not serve the long-term needs of the world population. Producing more fossil fuels to burn or logging old-growth forests for special metals or cattle ranches will only hasten climate and health catastrophes. Not coincidentally, many young people are speaking out about these global environmental issues and taking action. If they all without exception, from those at the center to supporters on the sidelines, would become vegan and promote veganism, the movement could boast yet more credibility.

Synergy, Genetic Control, and Kin

Rather than "cooperation" (a term I use along with collective action), one might consider how complex systems evolve through synergy. Peter Corning (1998) says synergy occurs when discrete pieces cooperate interdependently to produce a complex effect for the whole from various parts. Matthew Fisher (2019) covers supportive ecological interactions, generally, as well. In evolutionary terms, plants are not "decorations," says Karl Niklas (2016), but life forms that through flowering and seeding enabled other organisms to evolve. There's a holism where the end net result is different from what each individual element would produce on its own. Borrowing from this metaphor, we could modify the multiple animal agricultural structures toward a vegan economy. A shifting of the many pieces (cattle and dairy farms, packaging and production facilities, and grocers) could simplify output toward a green goal as part of cultural evolution. Right now the ultimate benefits of animal farming are chunks producing bad health and environmental degradation for profits at the expense of workers and sentient beings. With this industry where various companies compete with each other there's no meshing into a holistic, or healthy and sustainable economy. If meat and dairy farming are fractured, and if all the new units are vegan based, the composite result generated from the individual parts could become beneficial in terms of nutrition and ecology. In this effort, vegan agriculture could, at best, be more diverse and stratified for better social germination.

Synergies exist naturally in nature, as in evolved ecosystems, but not necessarily in the human realm where business and political competitiveness predominate and force imbalances. The measure of nature, all told, has always been fruitful sustainability, seen in Lovelock's (1979) Gaia hypothesis. Though biotic systems are discussed less theoretically, see, too, Matthew Fisher (2019). Within the past hundreds of years, however, human handiwork offers a much dirtier and less ecological calculation. For example, in the natural world for millennia water was somewhat evenly distributed. We could say it found its level, and those who required water gravitated toward it and shared the resource freely with other thirsty creatures. Water has now become a commodity. Rich countries have more water than some poor nations, and yet some corporate executives don't see access to water as a natural right. In the meantime, most modern industrial societies waste and contaminate water. Much water is required to produce one pound of beef, and that water ultimately becomes filthy runoff. Where synergy in nature evolved mutualistic structures (Lehman, et al. 2019), the human species has imposed dirty, artificial controls for its own gains that have essentially disrupted the ecologically balanced system.

In just over forty years beginning in 1970, say Elisabeth Bakker and Jens-Christian Svenning (2018), there's been a 58 percent decline in much terrestrial and marine wildlife. Such mass extermination comes at the hands of modern humans. Even if people don't care about animals, these beings, say the authors, provide critical eco-system functions that, once lost, offset the equilibrium in nature. This conclusion of compromised environmental health globally is verified by Randolph, et al. (2020). That impact of instability does not bode well for any living creature, man, woman, or child. Rewilding loss of flora and fauna could mitigate some of this severe eco-logical damage (Perino, et al. 2019; Broughton, et al. 2021), but that depends on cultural shifts that respect nonhuman species, insects, and foliage in their natural habitats. One suggestion might be to take the concept of trophic self-regulation or physiological energy in food webs (Fisher 2019). This idea could be applied to most modern societies in the form of veganism and local, supply-conserving organic veggie farms and plant food distributors. Reintroduction of eco-friendly species to nature along with the cessation of the high pollution content of meat and dairy farms can become a major step to mitigate poor health and climate change.

Genetic controls, intimates Jerram Brown (1975) in yet another standard text, are at the evolutionary basis of behavioral differences and even similarities among species. The evolution of behavior is roughly equivalent to behavioral biology, distinguishing the innateness of genes with instinct and learning through the environment (Clark, et al. 2018). This is most evident when comparing great apes to humans. Most people are fascinated by great apes since they look and behave like us on many levels. The differences are more striking. For instance, great apes tend to maintain sustainably their once vast, but now shrinking, forest habitats far better than most modern humans (Tague 2020). Brown says there are three main social behaviors from selection pressures.

1.   Aggression and distancing (individual selection).
2.   Mating and reproduction (sexual selection).
3.   Offspring care and rearing (kin selection).

For our purposes, let's place number 2 aside for now, but consider numbers 1 and 3.
Human evolution has favored adaptations for aggression, separation systems, and parental care (Clark, et al. 2018). Like individuals, groups are competitive, and

nations are aggressive toward each other. The Paris Agreement on climate change was a spectacular effort for almost two hundred nations to unite in determination to fight global warming. Yet under the Trump administration, the United States, a key player, pulled out (Shear & Davenport 2020), as that administration also threatened to leave NATO and defund itself from the World Health Organization. One could enumerate other, ecologically disruptive heads of states who jilt the balance of nature's economy. Such divisive behavior or nation-first thinking is not environmentally sustainable on a global level, and it's surprising there are no punitive measures by other countries against such an intimidating and dangerous frame of mind. In a synergistic mindset, all players lose in this scenario.

As for kin selection theory (see Clark, et al. 2018), the nation-first thinkers might assert that privilege, but since all humans are related genetically, and supposedly in an effort for guaranteeing future health, then nations cannot afford to stand alone or harm the shared oceans, forests, open plains, and atmosphere. Slashing and burning tropical or rain forests to fill overflowing corporate coffers deleteriously affects the health of people far away. Recall those Americans in Dearborn, Michigan. All nations might promote and assist people, as best they can, in a vegan diet. Currently, veganism is considered an ethical choice though some make that "choice" for health, celebrity, religious, or even economic reasons. That ethos can, with little effort, become more widespread. In terms of relative fitness, where one genotype might be more successful than another (Clark, et al. 2018), it matters little if the consuming instinct of humans helps them predominate at the expense of most other life. There is no real dispersion pattern among humans since with 8 billion hungry people we inhabit, and mostly pollute, all parts of the planet.

Perhaps famed biologist E.O. Wilson (2016) is right: half the earth should be left alone to biodiversity without people. Humans, too, need a biosphere. It's unlikely that people will abandon huge swaths of earth for rewilding with resource acquisitive nations vying over others in land grabs. Fortunately, at the One Planet Summit (2021) over fifty nations have committed to preserving 30 percent of their land or sea. Adam Vanbergen, et al. (2020) suggest that rather than expanding the already huge cattle ranches and livestock farms, the land could be reclaimed for better purposes, like cultivating bioenergy materials, sustainable timbers, carbon sequestration, as well as biodiversity reclamation. In other words, we can evolve newer behaviors to fashion a better future for all our human relatives.

From an evolutionary standpoint, taking into consideration collateral costs, it makes sense to invest in your stomach, offspring, mate, or kin. As with other species, for humans there is territorial behavior, literally and figuratively, in terms of a variety of plant, animal, and mineral resources, including other people and even mates. Among many animals, social behaviors, whether agonistic, sexual, or aid-giving, provide an advantage (Clark, et al. 2018). In my argument, let's not rule out the first two (aggression toward vegans or other states and meat as enhancing sexual favors), but attention is on the third. In biological language, aid comes in the form of care for young or kin and communal cooperation. Parental care evolved through natural selection to perpetuate one's own genes. Larger numbers of people in many industrialized nations now need to make the connection between environmental care and genetic longevity. Ecosystem engineers, from beavers to great apes, figuratively invest in their habitats as if through some version of kin selection (Tague 2020). Modern humans typically do not.

Fitness is relative (Clark, et al. 2018). This means that in a transaction one gains and one loses. While there is, in line with James Lovelock's (1979) notion of Gaia a self-regulating stability in nature, some purblind humans have dramatically tilted the equation to favor their fitness at the expense of others. Unfortunately, the plunder of natural resources and the disproportionate farming of meat, seafood, and dairy, as could be predicted with a world population of 8 billion, has not favored humans and is now a detriment to many life forms. Just as the broad term "environment" includes social aspects, humans, like other species, now need to be social toward the land, water, air, and animals in an ecological relationship where the outcome is mutual aid, as Peter Kropotkin (1902) hypothesized. We should not be surprised to see cooperative communities in nature, either through kin selection or reciprocal aid (Nowak, et al. 1995; Clark, et al. 2018). That was the case for hundreds of thousands of years (millions if we include australopiths) and is true today with hunter-gatherer communities, but not so with most industrial societies.

Jos Kramer and Joël Meunier (2016) opine that group and kin selection are not at odds but complement each other by utilizing optimum phenotypes in the long process of selection. Here, too, in this argument it's clear that veganism offers the best health and survival choice but requires multilevel selection to prosper and benefit large numbers. To be clear, lined up with Brown (1975), behavior in how most humans treat the environment is a consequence from a long trajectory of "accumulated interactions." Genetic heritability can function as a population factor (genotype) not to be confused with developmental learning (phenotype). This means we are not doomed to massive meat and dairy consumption but can learn how to be healthy on individual and population levels by accepting veganism.

Biota, Allometry, and Energy Budgets

Biogeography, as seen through the work of renowned scientist George Gaylord Simpson (1967), demonstrates how each organism has its own ecological niche. Other authors address microscopic, plant, and animal construction of place, such as Lisa Bartee, et al. (2017), Matthew Fisher (2019), and Mary Ann Clark, et al. (2018). An organism's geographical distribution is explainable, in part, by evolution and geology, and the diversity of species occurs over time via migration from land shifts and other environmental changes (Clark, et al. 2018). Jaw angularity, in particular, increases with herbivory (plant eating) while the length of the jaw's posterior decreases with herbivory, relates David Grossnickle (2020). This is because powerful muscles are needed to grind down plant matter. Furthermore, body mass might not simply be an equation of herbivory but is influenced by other non-feeding factors. Diversity of ecology can influence body mass, but other selection pressures can affect functional morphology, says Grossnickle. Jaw reduction in

Homo seems to have arisen from the ability to process and particularly soften food not so much with teeth but with tools and then fire. Incisors and mandible robustness, say Veneziano, et al. (2019), were reduced, but not necessarily the premolars or molars. Although human dentition will be discussed in more detail in Chapter 4, suffice it to say now that australopiths, our early ancestral relatives, exhibited larger chewing surfaces and thick enamel of premolars and molars, mainly *Paranthropus*, to eat herbaceous vegetation, roots, and tubers when forests diminished and grasslands increased. Grassland patches no more than 135 square meters would provide enough energy to keep a large body hominin satisfied for a day (van Casteren, et al. 2020). Veneziano admits that morphology might not always indicate diet and could be affected by body size and phylogenetic restrictions (i.e., hominins are among catarrhines, simians with a narrow nose).

Chiefly for sedentary species like trees, it's questionable as to whether any ecological niche is ideal. Rather, species must adapt to their surroundings (Wohlleben 2016). Humans, however, can live anywhere, but there have been ecological consequences of their subjugation of the earth. Like great apes, early hunter-gatherers knew where to find plants and insects for their food sources. This foraging strategy was abandoned and followed by sedentary agriculture (Clark, et al. 2018; Fisher 2019). Most, if not all, city dwelling modern humans do not know how to survive in a forest or how to farm vegetables. In part, this artificial city biogeography has created a false dependency on meat and animal products like dairy milk and cheese, as if they arise from the supermarket. There is no biotic interaction of plant and animal species in most modern cities where humans have become dependent not on the earth but on services provided by others. That could change if vast city wastelands were to become community gardens teeming with vegetable life.

However estranged most modern humans are from nature, even in small caches of urban brick and concrete there are biotas. Plants grow from between the crevices in chimney stones. Ants swarm in busy lines on playground concrete. Backyards teem with beetles, worms, bees, pollen, and birds. Biological changes from natural selection on the level of the organism can result in changes to the geographical evolution of the biotic community, says Simpson (1967). If a forest shrinks and becomes a savanna, fauna and flora must adapt or die. Life history is related to geography (Clark, et al. 2018; Fisher 2019). Humans are not alone in their expansion. Not counting contractions, of which there are some, life forms have essentially multiplied in diversity since the Cambrian explosion about 500mya. Other species could conceivably conquer the earth, and perhaps some forms of bacteria, viruses, or insects will. Organisms that share an environment and use similar resources will compete for dominance. So species adapt, evolve, and perhaps occupy other habitats. In terms of feeding, Simpson says when animals become specialist eaters there are more kinds of them (Clark, et al. 2018). That's not necessarily the case among great apes, including humans, essentially frugivores/folivores or fruit and plant eaters. People do not need to eat one type of food, but yet there is a false dependency on meat. This is not an exaggeration. During the 2020 pandemic, the president of the United States ordered meat packing plants to remain open, in spite of worker deaths or illness from the virus, implying meat as the only leading food source.

The filling of a biotic vacuum likely occurred with the demise of australopiths, the rise of early *Homo*, and eventually the dominance of earth by *Homo sapiens* who perhaps drove Neanderthals to extinction competing for the same space and similar resources. As we now do with great apes limited to reduced tropical areas, and as we perhaps did to Neanderthals, pursuant to Malthusian theory, modern humans wedge themselves into the habitats of other species in competition to exploit resources (Hodge & Radick 2003). Typically, ecosystems represent a balanced harmony, evidenced by the biodiversity of tropical and rain forests. Modern human beings, on the other hand, have for the most part supplanted or eradicated biodiverse areas, whether in European, North American, or Asian cities, with their own domineering populations.

Let's look at nature's biomes from a slightly different angle. Allometry, the concern with different measurements, is a biology of scaling and consists of the "consequences of size" (Calder 1984) related to an organism's life history in her ecological place. More recently, J. Gordon Betts, et al. (2017) address scale in environments, too. All of an animal's body parts and physical actions need to be considered as one form and function (Clark, et al. 2018). A large animal might require more food in mass, but a much smaller creature to scale needs as much or perhaps more food because of its higher metabolism (Calder 1984; Betts, et al., 2017). Through biogeography, there can be limits on an animal's size, shape, diffusion, and bioenergetics (Clark, et al. 2018). Excessive competition is inefficient and why there are differences in size, diets, and foraging strategies among species. If primarily vegetarian orangutans and gorillas are larger in bulk compared to chimpanzees, with a wild, adult male chimp potentially weighing 150 pounds, humans do not need lots of meat. That is, chimps desire meat but eat far less of it than humans, yet they live harder lives outdoors, need to climb and forage, endure heat stress and fasting. The standard argument against this thinking, of course, is that humans have a larger and more energy dependent brain. The brain matter will be addressed shortly. Without doubt, that human brain now needs to work more efficiently in solving the growing trend to obesity, other health problems, and vanishing green environments by looking to veganism. If humans are truly the wisest of species, they will dramatically alter their diet for their permanence and preservation of nature.

Great apes need calories for forest survival functions that have been eliminated from a human life history. Stressors on humans are mostly self-inflicted, though that evaluation takes into account most people in many industrial societies and not in developing countries. For instance, the worry is not about where the next meal will come from or the presence of a predator but how to multiply funds in the bank account or cars in the driveway. The external environment in which apes live is more indifferent and costly than the indoor environments of most modern humans. If fitness is survivability to reproduce, many species, including our near primate relatives, do so without a disproportionally daily intake of meat. Allometry, says

Calder (1984) in his standard text, consists of correlations not only of an organism's physiological characteristics but also adaptive strategies to its environment. Body growth and proportions are related to environment (Fisher 2019). In other words, an organism can survive to the maximum reach, consuming and using more energy, by conforming to an aggressive environment. Not to generalize too much, but if weighed, most modern human diets are not calorically scaled to their environments, and hence obesity, diabetes, and cardiovascular disease (Betts, et al. 2017).

In another established book, Robert Peters (1983) talks about the ecological implications of body size in terms of respiration, metabolism, physiology, locomotion, ingestion, growth and reproduction, mass flow (e.g., blood and urine), and nutritional turnover. More recently, Mary Ann Clark, et al. (2018) echo much of Peters in their discussion of animal structure and function, as does Matthew Fisher (2019). There are action dynamics in ranges scaled to body size. Peters tries to demonstrate how warm-blooded animals always need abundant resources where body size alone does not explain metabolic rates. His work confirms the principle of similitude. Similar organisms are parallel in measure regarding bio-physiological changes when metabolism is the sum total cost of processes in relation to one another. Compared to apes, humans are hypermetabolic, an evolved adaptation that worked well in prehistoric times and for hunter-gatherers (Pontzer, et al. 2016). For most modern humans, we overcompensate with this metabolism by eating too much, and usually the wrong foods.

Consider, says Peters (1983), loss of weight in relation to survival. No animal can live if it loses half the weight of its usual body mass. Fats in some animals can be lost during sleep. None of this explains the disproportion between our nearest living relatives, great apes, other primates, and the obesity and pollution rates of humans. If we are smarter than "animals," we don't show it since we've created such environmental chaos with our consuming instinct. As I demonstrate in *An Ape Ethic* (2020), great apes evolved adaptations of physiology, morphology, population size, diet, and socialization as ecosystem engineers. Most humans have not. In terms of the human organism's food ecology of meat, seafood, dairy, etc., there is no balance between function and structure (Betts, et al. 2017). Some of these comparisons will be useful to remember in reading the next chapters.

Importantly, there is little difference between the ingestion rate of carnivores or herbivores in how food becomes tissue and energy, Peters (1983) says. See, too, Mary Ann Clark, et al. (2018) and Matthew Fisher (2019) who explain related ideas. Herbivores and carnivores have different energy diets, whereas folivores assimilate food in such a way that they compensate by reducing activity or increasing food intake (e.g., Betts, et al. 2017). Seeds, on the other hand, assimilate in a higher degree than plant foods. What must be taken into consideration is net result. For instance, hunting for meat depletes more energy as does roaming far and wide for plants and so why some omnivores eat meat if it's readily available. One can gain more energy from foods not highly assimilated if the cost in sourcing them is not great. This is why zoo animals have low ingestion rates. Wild animals will on occasion increase ingestion rates to accumulate fats depending on growth.

So there's likely a correlation between human sedentary rates, animal agriculture, eating immoderation, and all of the consequences of those behaviors, like poor health, corpulence, and environmental degradation.

As for energy budgets, flesh and seeds achieve higher rates of assimilation than new vegetation and, even less so, mature vegetation, correlated to an organism's "mass flow" (Peters 1983; Clark, et al. 2018). For instance, says Peters, one study shows how primate population declines with an increase in individual body size and carnivory, a balance. The opposite, it seems, is true of humans. If chimp habitats are artificially decreased by humans but their carnivory continues, there will possibly be a net loss of monkeys they prey upon since density for all animal activity is a factor. Loss of foraging territory might even explain an increase in chimp carnivory, a response to food stress. Tropical species evolved to have a low population concentration compared with temperate species and why carnivores, like large felids, tend to have a lower population density. Economically balanced growth needs to be examined. Hunter-gatherers and nonhuman primates, in contrast to most modern human societies, have achieved a homeostatic sympathy with ecosystems (e.g., Betts, et al. 2017). Farming of meat and dairy has set the flow of nature off kilter, with an obvious increase in disease among both humans and animals as well as pollution, whether methane gas (Steinfeld 2006) or offal runoff contaminating water supplies and land.

Evolutionary Psychology

In his influential text, psychologist David Buss (2019) says that humans evolved and adapted in an organic climate that enabled them, cross-culturally, to understand how animals have a characteristic, not hierarchical, "essence" that, like their own, grants certain qualities and temperaments. That intuitive sensitivity has been lost in most modern humans who inhabit high-rise buildings in large cities. Buss talks about this perceptual sensation in terms of nutrition intake and toxin avoidance, but we can apply this acknowledgement of quintessence to other cross-cultural beliefs about living organisms. Many cultures believe in life forces or spirits. A problem with animal trafficking in endangered species is how some cultures believe parts of certain animals, whether the hands of gorillas or the tusks of rhinoceroses, hold certain magical powers. Of course they do not. The point is that via theories of cultural evolution and evolutionary psychology, visionary leaders, governmental policy makers, educators, and artists have the ability to influence and shift cultural norms, *mores*, and beliefs about and attitudes toward animals. These leaders could help shape a culture, primed by voices from the corners, toward vegan agriculture and environmental health as a new centerpiece.

Similar to chimpanzees, Buss (2019) says that food is involved in human courtship, rituals, alliances, status, reconciliation, and generally as a means of forming relationships. Food is not only sustenance to survive and biologically reproduce, it bonds people culturally. We don't need meat alone to bond with others, considering the ultimate effects of animal farming. Nor do we really need cow's milk in coffee or tea. Buss goes on to say that food metaphors are prevalent in languages, like the words "sweet" and "juicy" or the expression "hard to swallow." All of this reflects ancient hominin psychology, still part of the adapted psyche of modern humans. There is a natural repugnance or disgust emotion to items one would not consume. This gag reflex is cross-cultural. No one would eat maggot infested, rotting flesh or feces, though more will be said about cultural acceptance of rotten meat in a brief discussion of Neanderthals, Chapter 4.

Disgust in order to evade disease is not unique to humans and can be spread culturally. According to Trevor Case, et al. (2020), great apes also exhibit disgust aversion responses, although "muted." That emotional response could possibly be applied if animal slaughterhouses were on Main Street, U.S.A., with glass rather than brick walls. Paul Rozin, et al. (2008) have done extensive research concerning disgust responses that are evolutionary, powerful, and persistent. In listing the nine spheres of disgust for North Americans, not surprisingly Rozin places food first, probably because it's the most common substance with which, like our prehistoric ancestors and other animals, we have contact. One could go as far as saying meat eaters are politically disgusted with vegans whom they erroneously label as bleeding-heart liberals. Research by Rozin, et al. (1997) shows how moral vegetarians at first persuaded because of animal rights later became disgusted at the thought of eating meat. This is evidence of how beliefs can alter emotional states, crucial to the argument for a vegan culture and claims about our historically plant-based food ecology.

Any disgust reaction is of course an evolved adaptation to protect the body from disease and harmful substances, but the response is subject to cultural evolution. We might be able to duplicate that reaction as a response to animal cruel and environmentally degrading practices in meat and dairy farming. Culturally, humans have shielded themselves from the bloody spectacle of mechanized, mass animal slaughter. In Thomas Hardy's novel *Jude the Obscure*, Jude in the rural English countryside of the 1890s, who was told as a youngster to be kind to animals, finds he can't slaughter a pig by poking its neck vein and letting it die slowly. He cuts deeply and quickly to reduce the animal's suffering, but he's then told he has spoiled the meat. We have socially evolved a cognitive and ethical dissonance about eating animal parts, whether a cow's tongue, a calf's liver, a whole sardine, or a pig's foot. There are evolved, negative responses to bloody wounds with fear of disease or infection, but not when modern humans see packaged red meat. That acceptance of meat artificially colored, cut into neat squares, and wrapped in cellophane can be turned off and the disgust response turned on. Many vegans are disgusted at the thought of slaughtered animals for human food and at the sight of animal meat or organs displayed in supermarkets or hanging on hooks in butcher shops.

Buss (2019) says we can psychologically turn off the adaptation of disgust when we have to care for a sick loved one and, at that time, change adult diapers or dress oozing wounds. Intellectually, humans know from where the dinner meat on their plate comes. They avoid viscerally confronting the raw flesh by sanitizing it through cooking and the use of spices. From ages past it has been demonstrated across cultures that certain spices, like onion, garlic, oregano, and forms of black pepper act to kill microorganisms on meat. Salt would be used to preserve dead animal flesh and store it safely. Use of spices is likely a form of cultural acceptance of animal eating. People don't taste the meat, per se; protein has no taste. Rather, they taste the fat that has been seasoned. One could as easily enjoy seasoned, sautéed mushrooms or vegetables. Disgust responses about animal flesh are muted through psychological mechanisms and food flavoring, adaptive advantages. Let's remove the packaging and flavoring to grasp the emptiness of that preference.

Evolution of Diet

In *An Ape Ethic* (2020), I explicitly state how humans evolved from forest dwelling hominins and interrogate how poorly we now treat those lands in comparison with our living cousins, the great apes. The evolutionary case for veganism matters because if we know how our diet evolved, it can help us adapt to current selection pressures like climate change and zoonotic diseases. Humans regularly deforest tropical areas, encroach on diminishing animal habitats, and consume animal flesh, including endangered species, at alarming rates. (With hope, from environmental activist pressure, many governments will limit deforestation soon.) It matters not if you don't eat bats; someone across the globe eats those creatures, acquires a virus, and then takes a plane to your city. As our primate history demonstrates through an evolutionary reconstruction of diet, which will be examined in Chapter 4, humans are not born meat eaters.

You are what you eat, so if you consume lots of animal products, like fatty flesh and dairy, your intemperance will result in obesity and related ailments like diabetes, cardiovascular disease, high cholesterol, etc. The point is that where you live along with cultural pressures contribute to what you eat and in which amounts. In Cheryl Abbate's view (2020), meat eating is a product of cultural conditioning, as argued here. Regardless of our evolutionary origins, it is still highly problematic to consume other animals in contemporary society. Human decisions are made, mostly, in cultural contexts that presently glorify meat. In comparison, great apes like orangutans, mountain gorillas, chimpanzees, and bonobos thrive healthily from diets rich in mostly fruits and leaves. Plant leaves contain sugars, starches, proteins, and water. Humans have culturally evolved cooking and eating meat, advanced the use of dairy products through farming, but can progress away from these practices that, in the hands of multinational corporations, have become physically unhealthy and environmentally unstable. The conventional story is that our brain size increased through the cooking of meat, although that's a late development at around 500kya according to Zink and Lieberman (2016). Even so, at that time we were not anatomically modern humans.

Brain size is relative to fitness factors and habitat (Clark, et al. 2018). Likewise, there is solid science to dispute the notion of enlarged brains only through cooking meat. Neanderthals and human predecessors as far back as 600kya were eating lots of energy-rich carbohydrates from cooked starchy roots and nuts. These findings come from a study of oral bacteria (Fellows, et al. 2021). Meat, which is not high in glucose, was therefore not the main driver of a larger hominin brain. Some authors (Barr, et al. 2022) question the myth that by eating lots of meat *H. erectus* catapulted human anatomy and behavior, inferring instead that dietary variety was paramount. Alianda Cornélio, et al. (2016) show, using modeling similar to those arguing brain encephalization from cooking, that brain expansion is independent of fire control and heat processed flesh. See the brains of great apes, dolphins, and elephants, for instance. Alex DeCasien, et al. (2017) question the social brain hypothesis of enlargement and offer an alternative idea where brain size is determined by primate diets of fruits and leaves and not sociality.

Brains evolved to control a body in its ecosystem, not really to think (Barrett 2011). Whereas in most animals cognitive processes spread outward from the brain to connect and interact sustainably with the environment, for humans brains have been used to dominate biomes with destructive and wasteful results. Nevertheless, brain and neuron network expansion through social complexity is a valid factor. Before farming, ancient human species likely ate microgreens that have high concentrations of nutrients (Xiao, et al. 2012). With brains, size alone is not a determining factor. On average, Neanderthals had larger brains than ours. Many species evolved tiny brains and survive well. What matters more are the neural connections and how they are used. In an intriguing argument, philosopher Baptiste Morizot (2021) sees most human thought, whether in mathematics or philosophy, as an exaptation of early human tracking skills. The group collective evaluated the tracks, all spoke, and data was compared to posit a plan. Primatologists have posited great ape intelligence in their ability to identify, recall the locations of, and find hundreds of food items. What's different here is how Morizot does not lay emphasis on the hunting but on the tracking with skills like understanding the mind of another animal, prediction, anticipation, and self-control. Biological and cultural anthropologists have posited a number of theories concerning the development of intelligence, and reliance on one aspect neglects to pull them all together. While it's difficult to disagree with Morizot, we'd also have to consider the cognitive abilities of tracking along with, for instance, the mother–infant dyad, a social brain hypothesis in group living, so-called Machiavellian intelligence, and consciousness in material culture, to name a few.

While they don't claim to have answers about what has driven brain expansion in humans, since hypotheses from socialization, deception, and foraging are factors, Leslie Aiello and Peter Wheeler (1995) claim that brain increase and gut decrease correlate to eating animal products. Their focus is on, as a representative, the large brain/small gut of *Homo ergaster* whom they claim preyed on animals, compared to *Australopithecus afarensis*, as an illustrative species with a smaller brain. This simplistic physiological analysis ignores other factors of brain size and intelligence, including nutrition from plants, physical environment, family dynamics, etc. (Bartee, et al. 2017). There are wide variations in human brain size, anyway. The question is whether or not there was a direct correspondence between gut and brain. With rudimentary fire, early people likely processed vegetable foods to increase digestion and bring out nutrients. If modern humans with large brains are super intelligent beings, it's only logical that through cultural evolution we can nearly eliminate meat and dairy eating to achieve better health, control climate change, and protect animal lives in ecosystems. While I draw analogies between modern humans, australopiths, and living apes, of course there are differences. So-called higher quality food like meat did not inevitably drive human evolution, for there are many species with supposedly low-quality food who have highly advanced intelligence and complex social behaviors.

There are multiple dimensions of evolutionary processes along with a range of selection pressures for species, many of whom currently represent a relative peak in their evolved form. Aiello and Wheeler (1995) seem to be pushing a punctuated equilibrium moment where there is a spike from australopiths to *Homo*. Evolution through descent with modification is more gradual, but with cultural evolution we can effectuate much needed social reforms somewhat quickly. Many factors would have contributed to brain encephalization. Trying neatly to unify any number of selection pressures into a single cause for a large brain can lead to strenuous claims. For instance, Miki Ben-Dor and Ran Barkai (2021) say that human brain size and other behaviors evolved as adaptations to hunting small prey after large animals had been hunted to extinction. With that line of thinking, some chimpanzee groups who are more often hunting colobus monkeys (Chapter 3) will evolve into a super species. Yet others (Benito-Kwiecinski, et al. 2021) attribute the large size of human brains to cell shape as a prime force, as if multiple selection pressures are not in control. Mutations alone won't determine evolution, since selection pressures must act on those alterations. That's the crux of this whole argument: cultural selection can help drive positive changes through dietary modifications.

The argument on the biological side of the evolutionary case for veganism need not be airtight. A review of the literature shows much discussion and dispute. Living nonhuman apes eat insects and at times meat (Chapter 3). Early humans scavenged carcasses, then hunted small game, large game, fished, and eventually farmed animals for food (Chapter 4).

Flowing from this completed chapter, upcoming points in the bulk of subsequent analyses are as follows.

1.   Our living ape cousins still rely mostly on fruit and leaves.
2.   Deep history shows our hominid ancestors as primarily fruit, leaf, and root eaters.
3.   As omnivores we have evolved all forms of plant and animal agriculture.
4.   Therefore, we can culturally gravitate to a vegan economy in order to increase human health, sustain biodiversity on the planet, and stave off the most deleterious effects of climate change.

These key points will take us into the next immediate chapters about great apes and early humans. Just as we are biological animals dependent on food and ecology, so too are our nearest living cousins, the great apes. Looking at their nutritional bionetworks sheds light on us; further inspection comes in Chapter 4 as we peer into the distant past and examine the nutritional ecology of our hominin ancestral relatives. All of this will tie together in a discussion of cultural evolution in Chapter 5 as we plan our future food ecology on planet earth.

References

Abbate, Cheryl. 2020. "Meat Eating and Moral Responsibility." *Utilitas* 32 (4): 398–415. https://doi.org/10.1017/S0953820820000072.

Aiello, Leslie C. and Peter Wheeler. 1995. "The Expensive-tissue Hypothesis: The Brain and the Digestive System in Human and Primate Evolution." *Current Anthropology* 36 (2): 199–221.

Bakker, Elisabeth S. and Jens-Christian Svenning. 2018. "Tropic Rewilding: Impact on Ecosystems Under Global Change." *Philosophical Transactions of the Royal Society B* 373: 20170432. http://dx.doi.org/10.1098/rstb.2017.0432.

Barr, W. Andrew, et al. 2022. "No Sustained Increase in Zooarchaeological Evidence for Carnivory after the Appearance of Homo erectus." *PNAS* 119(5): e2115540119. https://doi.org/10.1073/pnas.2115540119.

Barrett, Louise. 2011. *Beyond the Brain: How Body and Environment Shape Animal and Human Minds*. Princeton: PUP.

Bartee, Lisa, et al. 2017. *Principles of Biology*. Open Oregon Educational Resources/OpenStax.

Ben-Dor, Miki and Ran Barkai. 2021. "Prey Size Decline as a Unifying Ecological Selecting Agent in Pleistocene Human Evolution." *Quaternary* 4 (27). https://doi.org/10.3390/quat4010007.

Benito-Kwiecinski, Silvia, et al. 2021. "An Early Cell Shape Transition Drives Evolutionary Expansion of the Human Forebrain." *Cell* 184: 1–19. https://doi.org/10.1016/j.cell.2021.02.050.

Betts, J. Gordon, et al. 2017. *Anatomy and Physiology*. Houston, TX: OpenStax/Rice University.

Blute, Marion. 2010. *Darwinian Sociocultural Evolution*. Cambridge: Cambridge UP.

Broughton, Richard K., et al. 2021. "Long-term Woodland Restoration on Lowland Farmland Through Passive Rewilding." *Plos One* 16(6): e0252466. https://doi.org/10.1371/journal.pone.0252466.

Brown, Jerram L. 1975. *The Evolution of Behavior*. NY: W.W. Norton.

Buss, David M. 2019. *Evolutionary Psychology: The New Science of the Mind*. Sixth edition. NY: Routledge.

Calder, William A. 1984. *Size, Function, and Life History*. Cambridge, MA: Harvard UP.

Case, Trevor I., et al. 2020. "The Animal Origins of Disgust: Reports of Basic Disgust in Nonhuman Great Apes." *Evolutionary Behavioral Sciences* 14 (3): 231–260. doi:10.1037/ebs0000175.

Clark, Mary Ann, et al. 2018. *Biology 2e*. Houston, TX: OpenStax/Rice University.

Cornélio, Alianda, et al. 2016. "Human Brain Expansion During Evolution is Independent of Fire Control and Cooking." *Frontiers in Neuroscience* 10: 167. doi:10.3389/fnins.2016.00167.

Corning, Peter A. 1998. "The Synergism Hypothesis: On the Concept of Synergy and Its Role in the Evolution of Complex Systems." *Journal of Social and Evolutionary Systems* 21 (2): 133–172.

Cosmides, Leda and John Tooby. 1992. "Cognitive Adaptations for Social Exchange." *The Adapted Mind: Evolutionary Psychology and the Generation of Culture*. Jerome Barkow, et al., eds. NY: Oxford UP. 163–228.

Darwin, Charles. 1839. Journal of Researches. James A. Secord, ed., *Charles Darwin: Evolutionary Writings*. Oxford: OUP, 2008.

Darwin, Charles. 1859. *On the Origin of Species*. Joseph Carroll, ed. Ontario, CN: Broadview P. 2003.

DeCasien, Alex R., et al. 2017. "Primate Brain Size is Predicted by Diet But Not Sociality." *Nature Ecology and Evolution* 1 (0112). doi:10.1038/s41559-017-0112.

Distin, Kate. 2011. *Cultural Evolution*. Cambridge: Cambridge UP.

Fellows, James A., et al. 2021. "The Evolution and Changing Ecology of the African Hominid Oral Microbiome." *PNAS* 118 (20): e2021655118. https://doi.org/10.1073/pnas.2021655118.

Fisher, Matthew. 2019. *Environmental Biology*. Open Oregon Educational Resources/OpenStax.

Grant, Verne. 1963. *The Origin of Adaptations*. NY: Columbia UP.

Grossnickle, David M. 2020. "Feeding Ecology Has a Stronger Evolutionary Influence on Functional Morphology than Body Mass in Mammals." *Evolution*. doi:10.1111/evo.13929.

Hatfield, Gary. 2013. "Introduction: Evolution of Mind, Brain, and Culture." *Evolution of Mind, Brain, and Culture*. Gary Hatfield and Holly Pittman, eds. Philadelphia, PA: U of Pennsylvania Museum of Archaeology and Anthropology. 1–44.

Hodge, Jonathan and Gregory Radick. 2003. "Introduction." *The Cambridge Companion to Darwin*. Jonathan Hodge and Gregory Radick, eds. Cambridge: CUP. 1–14.

Jablonka, Eva and Marion J. Lamb. 2005. *Evolution in Four Dimensions: Genetic, Epigenetic, Behavioral, and Symbolic Variation in the History of Life*. Cambridge, MA: MIT P.

Kramer, Jos and Joël Meunier. 2016. "Kin and Multilevel Selection in Social Evolution: A Never-ending Controversy?" *F1000 Research* 5: F1000 Faculty Rev.-776. https://doi.org/10.12688/f1000research.8018.1.

Kropotkin, Peter, 1902. *Mutual Aid: A Factor of Evolution*. NY: McClure Phillips.

Lehman, Clarence, et al. 2019. *Quantitative Ecology*. Minneapolis, MN: University of Minnesota Libraries Publishing.

Lovelock, James. 1979. *Gaia: A New Look at Life on Earth*. Oxford: OUP, 2000.

Morizot, Baptiste. 2021. *On the Animal Trail*. Andrew Brown, trans. Cambridge, UK: Polity Press.

Niklas, Karl J. 2016. *Plant Evolution: An Introduction to the History of Life*. Chicago: U Chicago P.

Nowak, Martin A., et al. 1995. "The Arithmetics of Mutual Aid." *Scientific American* 272: 76–81.

One Planet Summit. 2021. www.oneplanetsummit.fr/en.

Perino, Andrea, et al. 2019. "Rewilding Complex Ecosystems." *Science* 364 (6438): eaav5570. doi:10.1126/science.aav5570.

Peters, Robert Henry. 1983. *The Ecological Implications of Body Size*. Cambridge: Cambridge UP.

Physicians Committee for Responsible Medicine. 2020. www.pcrm.org/news/blog/president-trump-meat-not-essential.

Pontzer, Herman, et al. 2016. "Metabolic Acceleration and the Evolution of Human Brain Size and Life History." *Nature* 533: 390–392. https://doi.org/10.1038/nature17654.

Randolph, Delia Grace, et al. 2020. *Preventing the Next Pandemic: Zoonotic Diseases and How to Break the Chain of Transmission*. Nairobi, Kenya: United Nations Environment Programme.

Rozin, P., et al. 1997. "Moralization and Becoming Vegetarian: The Transformation of Preferences into Values and the Recruitment of Disgust." *Psychological Science* 8: 67–73.

Rozin, P., Haidt, J., and McCauley, C. R. 2008. "Disgust." *Handbook of Emotions*. Third edition. M. Lewis, J. M. Haviland-Jones and L. F. Barrett, eds. NY: Guilford Press. 757–776.

Ruse, Michael and Edward O. Wilson. 1985. "The Evolution of Ethics." *New Scientist* 108 (1478): 50–52.

Shear, Michael D. and Coral Davenport. 2020. "In Visiting a Charred California Trump Confronts a Science Reality He Denies." *The New York Times*13 September. www.nytimes.com/2020/09/13/us/politics/california-fires-trump-climate-change.html.

Simpson, George Gaylord. 1967. *The Geography of Evolution.* NY: Capricorn Books.

Steinfeld, Henning. 2006. "Livestocks Long Shadow. Food and Agriculture Organization of the United Nations." www.fao.org/docrep/010/a0701e/a0701e00.htm.

Tague, Gregory F. 2014. *Making Mind: Moral Sense and Consciousness in Philosophy, Science, and Literature.* Amsterdam: Rodopi.

Tague, Gregory F. 2020. *An Ape Ethic and the Question of Personhood.* Lanham, MD: Lexington Books.

Vanbergen, Adam J., et al. 2020. "Transformation of Agricultural Landscapes in the Anthropocene: Nature's Contributions to People, Agriculture and Food Security." *Advances in Ecological Research.* David A. Bohan and Adam J. Vanbergen, eds. Academic Press. 193–253.

van Casteren, Adam, et al. 2020. "Hard Plant Tissues Do Not Contribute Meaningfully to Dental Microwear: Evolutionary Implications." *Scientific Reports* 10: 582. https://doi.org/10.1038/s41598-019-57403-w.

Veneziano, A., et al. 2019. "The Functional Significance of Dental and Mandibular Reduction in Homo: A Catarrhine Perspective." *American Journal of Primatology* 81: e22953. https://doi.org/10.1002/ajp.22953.

Williams, George C. 1966. *Adaptation and Natural Selection.* Princeton: Princeton UP, 1996.

Wilson, Edward O. 2016. *Half-Earth: Our Planet's Fight for Life.* NY: Liveright Publishing.

Wohlleben, Peter. 2016. *The Hidden Life of Trees.* London: William Collins.

Xiao, Zhenlei, et al. 2012. "Assessment of Vitamin and Carotenoid Concentrations of Emerging Food Products: Edible Microgreens." *Journal of Agricultural and Food Chemistry* 60 (31): 7644–7651. https://doi.org/10.1021/jf300459b.

Zink, Katherine D. and Daniel E. Lieberman. 2016. "Impact of Meat and Lower Paleolithic Food Processing Techniques on Chewing in Humans." *Nature* 531: 500–503. https://doi.org/10.1038/nature16990.

3
GREAT APES AND OTHER PRIMATES

This chapter provides critical claims for the extended argument. Humans evolved from a deeply complicated history of primates, and, more specifically, share common ancestry with great apes. If you don't accept the scientific facts about human descent on ideological grounds or through some prejudice, then none of the rest of this book will make sense. In this writing, the main concern is about the ecology and culture of food. Great ape cognition, intelligence, and social behaviors related to their habitats are covered in *An Ape Ethic* (Tague 2020; see also, Herzfeld 2017). Consequently, though outside the scope of this writing, suffice it to say that genetically, morphologically, physiologically, and behaviorally humans are similar to great apes. In fact, we are apes. Michael Wilson (2021) believes there are adaptive peaks. For gorillas, it's body size and guts for fermentation. For bonobos, it's the moist forest. For chimpanzees, it's foraging in broad areas with culture. For orangutans, it's an arboreal life feeding from fruit trees. Presumably, we are to accept that the adaptive peak for humans is in the farming of animals that is driving poor health and climate change. As Peter Andrews and R.J. Johnson (2019) fear, we are developing physiological and psychological adaptations to eat highly processed supermarket foods. In the long run, that's not sustainable.

In order to survive, all animals must eat and find new and better ways to harvest food, minerals, and water. Wooden implements would have been easy to fashion, and still are by extant living apes. Stone implements found in Northern Kenya have been dated to 3.3mya, so before *Homo*, and perhaps as a result of many hominin adaptations to establish a wider dietary niche, say Rhonda Quinn, et al. (2020). Incrementally, early stone tools could have been used to process plant foods efficiently as well as animal resources, if any. Random, accessible stones employed for digging, scraping, puncturing, or peeling roots could have been used even before any deliberate fashioning of stone tools. Human behavior could be constrained by culture in how food preferences of a group can counteract resources in

DOI: 10.4324/9781003289814-4

spite of availability (e.g., a perceived need for meat over plentiful plant foods). Thurston Hicks, et al. (2020) talk about this relationship between tool technology and resources in terms of chimpanzees. Insect resourcing can differ in tool use culturally, but not conclusively be affected by ecological factors.

Diet is correlated to habitat and the consequential structures of morphology, like dentition, and physiology, like digestion and behaviors to survive and reproduce. Echoing earlier researchers, John Oates (1987) says that primate diets are predominantly eclectic, where eating a combination of fruits, leaves, insects, seeds, nuts, gum, flowers, bark, and roots is the best game plan. Karin Jaffe (2019) confirms what Oates says since, not surprisingly, ape feeding and diets have not substantially changed in the past generation or so. Earlier researchers have classified primates into three main categories: those who eat animal material or faunivores (hereafter insectivores); those who eat mostly fruit or frugivores; and those who primarily eat leaves or folivores. For convenience, let's separate insects from "meat," the latter of which can be defined as animal flesh. While some biologists will classify insects as meat, that's not the case here since modern humans eat a wide variety of animal flesh on a daily basis, not insects. For humans, meat means the skin, organs, bone marrow, and flesh of an animal, not insects. As a reminder, the term vegetarian is used to include a primate, as well as humans, who could occasionally eat insects, but typically not animals like birds, fish, or mammals. In primate origins, Keegan Selig, et al. (2019) suggest high insectivory, though some primate taxa are more adaptively evolved in a model favoring frugivory.

In *An Ape Ethic* (Tague 2020), a considerable amount of time is spent reviewing the intellectual abilities and behaviors of great apes related to their diets. The conclusion is that they are moral individuals by virtue of their land ethic as ecosystem engineers. Apes "care" for the environment to which they adapted. Their habitats, shelters, social settings, and food resources are attended to in thoughtful ways as if kin. The argument in that book is to grant apes citizen sovereignty over the lands they've inhabited for millennia and so ably maintain. On the other hand, modern humans have generally not cared for the environment and its many biodiverse species. Instead, we raise billions of animals every year, at great stress on and loss to environmental systems, only to slaughter and eat them. We have depleted sea foods from fresh and salt waters, simultaneously soiling those natural networks. Our actions are not those of ecosystem engineers and, worse still, our consuming lifestyles are driving many of our friends and family members to obesity, diabetes, and cardiovascular disease. Some apes confined in zoos, without the proper physical or social exercise, can succumb to the same health problems, like obesity, as many people in wealthy industrialized nations. This is not a judgment but merely an echo of the data (e.g., Hales, et al. 2020; Boonpor, et al. 2021).

A vegetarian or even vegan-like culture is part of our prehistoric roots, as shown in this chapter and the next, and should once again be embraced for all the good reasons previously enumerated in the opening pages of this text. Let's start, in this chapter, with the hominids (great apes), and in the next chapter deal with early hominins (humans).

Food for Animals

Food can be a constraint on primate populations because of scarcity, diffusion of nutrition, hard shells, or toxicity. If some primates, among monkeys, have evolved digestion to cope with foods like leaves, perhaps their populations are evolved adaptions. Compared to Old World monkeys, great apes need wider ranges with more ripe fruit. As a consequence, without having evolved a special digestive system, most great apes can eat non-fruit food items. Drawing from Juichi Yamagiwa (2004), it appears female apes maximize food resources for themselves and dependent offspring. Absent meat, mothers will share their breast milk and then other foods with offspring, help infants open or process food, and teach young over an extended period how to forage and find the best foods (Wilson 2021). Meantime, male chimpanzees will use meat for reproductive success, either by sharing with females for sex or with males to forge alliances (Pobiner 2020). Borrowing from Robert Trivers (1972), this type of "parental investment" might explain how among humans hunting and meat eating evolved, at least among females. Note, however, that gorillas and orangutans don't typically share food. To explain this difference orangutans are semi-solitary, especially males, and males don't form male–male bonds. Gorillas can be passive, and there is no true male–male alliance there, either. Among bonobos, the social structure is female centered, and foods are readily shared. While humans have a close genetic relationship to all great apes, it is particularly strong with chimpanzees and bonobos.

Peter Andrews (1985) notes how some views before the end of the late twentieth century perceived apes as "evolutionary failures" because there are so few of them and how they are geographically restricted. However, as demonstrated in *An Ape Ethic* (Tague 2020), these cited demerits are actually virtues. Because of their low and infrequent birthing intervals and their concentration in tropical or rain forests as seed dispersers, great apes are the true ecosystem engineers sustaining the lungs of the earth's atmosphere. Likewise, contemporary hunter-gatherers evolved systems to keep their needs in check with surrounding resources and, like the orangutan people of the forest, don't farm animals. On the other hand, the so-called merits of *Homo sapiens*, so many and widespread, are the causes of environmental degradation, typically in the form of farming cows, pigs, sheep, chickens, fish, and dairy, to name only a few "products" businesses foist onto hungry but well-fed consumers.

Talking about the ancient environment of evolutionary adaptation is not completely abstract. Jean-Jacques Petter and François Desbordes (2013) point out that in some parts of Africa and Madagascar we can still witness what they call an evolutionary laboratory. In other words, species descended from early primates, and a reptilian creature before that, continue to survive on diets of fruits, leaves, insects, etc. Every aspect of a tropical or rain forest is utilized by some species of primate for foraging or protection, from the tree-top canopies to ground vegetation. Ecological niche construction exists in abundance and, through selection pressures over time, has been employed advantageously. When a species carves out its own niche, there's less food and resource competition that, in turn, permits survivability and consequently morphological and behavioral variation over time. For example, say Petter and Desbordes, there can be any number of primate species in one locale of Asian or African forests, each performing life functions, nocturnally or diurnally, in a slightly different way. This is an illustration of species diversification. These habitats are also filled with other species that have evolved and differentiated in similar ways to live together. Humans, on the other hand, mostly tend to crowd out or obliterate flora and fauna that's in the way of their urban or suburban sprawl.

Note that the word forest is used loosely to include tropical forests of Africa and rain forests of Indonesia and the Amazon regions. There can be a primary rain forest, a secondary rain forest (where gaps created by fallen trees permit light and new growth), gallery forests (isolated), woodlands (with shorter trees), and savanna (sparse trees). Moreover, says John Fleagle (1988), within each of these forests, primates and other animals will occupy an ecological niche in which to survive and reproduce. Mary Ann Clark, et al. (2018) and Karin Jaffe (2019) discuss the importance and evolution of the ecological niche in terms of competitive exclusion. For instance, some primates might flourish at tree canopies where others have a life history lower down on trees or closer to the forest floor. Furthermore, different trees might be more conducive to some primates than others, depending on the preferred food choice. Key here is the diversity of evolved species adapted to micro ecological systems among thousands of plant species. Consider how modern humans don't any longer fit into but instead have forced themselves onto these structures coordinated by other species.

As Fleagle (1988) points out in his classic text, fruity plants evolved along with primates and through natural selection gained bright colors and juiciness ensuring edibility for seed dispersal. These ideas are highlighted, too, in Karin Jaffe (2019). Some plants evolved hard shells to protect their inner cores; consequently, primates evolved strong manipulation or dentition, thick enamel, and in some cases tools to secure the fruit. Wariness to carnivorous predators, like large felids (lions, tigers, leopards, etc.), helped evolve primate behavior gravitate upward into fruit trees. Primates who eat mostly insects because of their tree life rely on the creatures of those areas and have evolved adaptations of grasping hands and slicing teeth to better resource and process insects. Teeth reflect phylogeny, genetic control, and diet. Selection pressures affect dental morphology. In comparative studies with great apes and South African hominins, say Michael Berthaume, et al. (2020), tooth sharpness could indicate a high intake of plant fiber or some animal matter. Interspecific food competition between ancient hominins could account for differences in molar shape. Much more about hominin dentition will be said in Chapter 4.

Fruit is consumed early in the day for its sugar while leaves come later since leafy plants tend to have higher sugar content after a day of photosynthesis. A foraging strategy is such since it involves a complex array of costly and beneficial behaviors to maximize energy intake. To say fruit is favored simplifies how in tropical forests not all fruits come in bunches but are sometimes sparsely scattered. This means decisions about competition and even the risk of predation are factored into when and where to eat. Fruit is high in calories but low in protein, the latter of which comes from leaves and insects. Behavioral ecology and the different sizes of primates correlate, says Fleagle (1988). Primate behavioral ecology related to preferred plant and fruit variety with meat supplements among chimpanzees is studied in great detail by Karen Strier (2017). Gorillas are large and heavy but feed mostly on leaves. Smaller primates need more energy, have smaller hands, and can get all the nutrients, vitamins, and protein they need by eating several dozen insects per day. This might explain why larger primates rely on easily obtainable leaves and fruits supplemented by insects.

Behavioral Ecology and Bio-ecology

Man the hunter theory claims hunting is a bio-behavioral adaptation for our species (Lee & DeVore 1968). That's an antiquated idea; the future is in vegan agriculture. Human behavioral ecology asserts how flexible thinking and action enabled adaptations to different and changing environments. Cognitive aspects disclose that humans can see or predict consequences and respond advantageously, says Kim Sterelny (2013). If this theory is true from the long Pleistocene, we need to act adaptively now on individual, group, community, and national levels. Part of the problem is that humans, mainly with technology from the last few hundred years, have been engineering ecosystems to suit their needs and desires alone. We have engendered a sense of separation from and have lost the feeling of cooperation with the wild environment, necessary for survival evident in other species. Many humans are experiencing fitness decline, rising rates of obesity, diabetes, and cardiovascular disease. All of this is complicated with pandemic viruses of our own creation by deforestation and abuse of wildlife (Randolph, et al. 2020). Contrary to what others might say, based on clear evidence, humans at this stage of development have not optimized any environment, surely not the great outdoors, if all systems are failing. We need a new food model.

Behavioral ecology might also explain the expansion of meat consumption by humans over time, for if it were not meat and only insects we'd have had to eat lots of bugs. Even so, for maximum energy, protein from meat alone is not sufficient, and so why great apes with larger guts rely on fruits and leaves, the latter of which supply protein and carbohydrates. Folivorous primates tend to weigh no less than 500 grams, says Fleagle (1988), and insectivores weigh no more than 500 grams. Karin Jaffe (2019) discusses primate body size in terms of diet with similar conclusions as laid out here. Dentition is relative to bio-ecology as well, in terms of hands and nails, or claw-like nails, for manipulating a food source relative to teeth. Frugivores, generally, have larger incisors, compared to their molars, than folivores, based on how food structures are processed in the mouth. For example, cusps on molars to shear foods in folivores and insectivores tend to be sharp. Frugivores have more rounded molars to pulp out fruit substances. Primates who rely on nuts and seeds have nearly flat molars and thicker enamel. Look at your back molars in a mirror and see how they are mostly flat and wide for grinding plant foods, not for eating meat.

Variation in body size is evident in an adaptive branching away from a common ancestor. In this array, says Sue Parker (1999), a smaller-sized body (chimpanzees) employs an omnivorous diet while a larger-sized body (gorillas) leans to a low-grade herbivorous diet. Humans tend to fall somewhere in between. Body size for diet might be an evolved adaptation to limit food competition in a shared ecosystem, and this is probably true whether among living apes or extinct australopiths. With interspecific allometry among related species, some features scale up to body volume, like molars, while other features might not, like brains (Strier 2017). Even for primates who mainly eat insects, those same insects are feeding off plants. Epiphytes are non-parasitic plants that attach to rain forest trees; liana vines also attach to such trees. Dense trees with a canopy, along with epiphytes and lianas, provide a means of arboreal transport for primates as well as abundant, but widely dispersed, foods. Some trees fruit and some plants flower at times when others do not. Though somewhat stable, in spite of deforestation and climate change, there are dry periods and times when prized foods for some species are scarce.

Primates awake at daylight hungry and will consume high energy foods like fruits or tender leaves. Then, depending on the primate species and location, some forage and some delve into banquets of leaves. Protein-rich insects can follow the first feeding, but not more leaves since the early ones eaten must first be digested. Depending on the species and food, a good amount of time could be spent on ingesting and chewing. Like humans, primates need to allocate other parts of their day for resting and socializing, but this too varies across species based on feeding foods and foraging patterns, says Oates (1987; Strier 2017; Jaffe 2019), where insectivores forage longer than folivores who can rest more. The bottom line is that many primate species are selected for an intermediate position between strict folivory and insectivory. They are not dedicated meat eaters. A few insects would hardly qualify as "meat" for an ape or human. Regardless, and depending on location, some chimps eat few to no insects while at other locales many insects are eaten, determined by the time of year (Rothman, et al. 2014).

Food Distribution and Eating Behaviors

In the Malay Peninsula, says Oates (1987), throughout the day gibbons combine eating fruit and insects in various proportions. This routine indicates how primates practice energy economy in terms of accessing the distribution of foods (Strier 2017; Jaffe 2019). If a folivore finds food in seasonally short supply, activity is decreased. On the other hand, if a frugivore finds ripe fruits in season, he increases activity. Or, absent fruit, some primates will escalate time looking for insects or nuts. This is relative to species and location, and in competition there could be bouts of synchronized feeding. In some cases, again depending on species and group size, according to Oates, there could be interference competition with interaction or exploitation competition where one individual depletes most of a resource. Synchronization feeding, subject to group size, can increase interference competition, and hence why many species forage at a distance from each other. Yet, both these forms of competition can be replaced by cooperation when a primate group denies a food resource to interlopers.

Over evolutionary time, the number and dimensions of teeth (incisors, canines, premolars, molars) in primates and other species have changed because of various selection pressures to food sources and a changing environment. In terms of Old World monkeys and humans, there are incisors, canines, premolars, and molars on the top and lower jaw. New World monkeys have three premolars, one reduced to such a degree that it's lost in marmosets and tamarins. Apes have blunt molars for crushing fruit, and monkeys, like the colobus, have teeth specifically for shearing leaves. Apes have larger incisors and canines than humans, and in chimpanzees, orangutans, and gorillas canines of males are used for defensive purposes or fighting for access to females, not really for eating. Chimpanzees and bonobos prefer ripe fruit in trees. Chimpanzees overlap with gorillas in feeding sites, whereas bonobos have no such competition, even if friendly. Bonobos can eat as many leaves and shoots as they like for fallback food, though these are the preferred foods of gorillas. Some chimpanzee groups can consume hundreds of pounds of meat per year, but this is accomplished mostly through opportunistic hunting. It should be emphasized that chimpanzees do not use meat solely for food. Rather, meat is shared for sexual favors from females or for status alliances with males, so it's like a commodity. Besides, the "hunting" and sharing behaviors of chimpanzees can vary from group to group and place to place.

Gorillas eat mostly shoots and leaves, a fairly stable resource, but will eat fruit if it's available. Compare this to chimpanzees and orangutans who rely on fruits. Gorillas seldom eat meat, if at all. Schaller (1964) says he never saw wild gorillas eat animals, even given the opportunity of coming upon a bird's nest with eggs or a newly dead duiker. In zoo captivity gorillas can be encultured to eat meat, he adds, in line with the argument here about the cultural ecology of food. So meat eating is not absolutely necessary for humans if so for our great ape, and especially gorilla, cousins who closely share our DNA, morphology, physiology, and social behaviors. Some might claim that's an unfair analogy since great apes have evolved guts different from humans. Of course, that's partly the point. As omnivores, there are only good reasons to preserve health and the environment by reverting to a plant-based diet.

Orangutans have thicker dental enamel than chimpanzees or gorillas, mostly because they prefer hard-coated fruits and seldom eat meat, if at all. Gibbons eat ripe fruit, supplemented with bird eggs and insects. Siamangs eat fresh leaves and shoots, supplemented with fruit and insects. Old World monkeys, such as Cercopithecines, eat fruits (ripe and less so) and possess blunt molars to crush fruit. Other Old World monkeys, like Colobines, eat leaves and have molar crests to shear leaves. New World monkeys, Ceboids such as squirrel, capuchin, and owl monkeys, prefer fruits, nuts, and leaves, supplemented by insects. Some of these latter species, like tamarins and marmosets, feed mostly on insects and gum. Tarsiers in

Asia are nocturnal and eat mostly insects, supplemented with small vertebrates and then vegetation. Some species of lorises (of Africa and southern Asia) rely on insects, gum, and fruits. Species of lemurs (from Madagascar) will eat insects and gum, others eat leaves or fruit. Evidently in our primate and ape cousins, insect "meat" is only a small part of the diet, if at all. Chimpanzees and to a lesser extent bonobos pose a wrinkle, but we'll get to that shortly.

According to anthropologists Robert Boyd and Joan Silk (2017), primates have a good amount of dexterity in their hands, which can be used to catch insects or other prey. However, that's not necessarily why hands evolved, since primates are mostly arboreal and required reliable grasping ability for branches. Occasional arboreal behavior would have been part of our australopithecine relatives, too. Most primates prefer leaves, fruit, and flowers not fully mature since they lack toxic or unpalatable alkaloids. One can hypothesize that in eating leaves insects were consumed, at first, accidentally and then with purpose. From some common, distant insectivorous ancestor the muted habit remained (Klein 2009; Clark, et al. 2018). Perhaps not coincidentally, insects are eaten in some human cultures to this day.

In terms of food and energy, there are different types of metabolism, whether resting, activity, growing, or reproductive and nursing efforts for females. Regarding density and availability in tropical forests where most primates live, mature leaves are more abundant yearly as opposed to fruits, flowers, or insects. This accounts for dietary and range specialization across primates and great apes where folivores tend to have smaller ranges for feeding and nesting. Important primate nutrient sources, following Boyd and Silk (2017), are as follows.

1.   Leaves, seeds, stems, gum, roots and tubers, and insects (including mature leaves if digestible) for protein.
2.   Fruit, flowers, sap, tubers and roots, (including insects and gum if digestible) for carbohydrates.
3.   Seeds and insects for fats.
4.   Seeds, immature leaves, and insects for vitamins.
5.   Immature leaves, sap, tubers and roots, or insects for water.

Normally, insectivores are smaller than frugivores who are smaller than folivores, so body size and diet are related (Clark, et al. 2018). The smaller primates seem to require more energy foods, like gum and insects. Larger primates don't seem to need as much time to process their foods. Teeth are similar in great apes and humans, but differ from other primates. If, for example, a primate relies on gum, incisors are more pronounced for scraping and digging into trees. Cusp height and sharpness in molars tend to be greater for insectivores and then for those who feed on leaves. Rounder and flatter cusps are in frugivores. Very heavy enamel appears on teeth of primates, including humans, feeding mostly on hard foods like nuts and seeds. In terms of digestive systems, insectivores have a short and simple gut for maximum absorption. Frugivores can have a simple digestive system but a large stomach to hold a good amount of fruit and leaves. Because they need to digest lots of cellulose and secondary plant compounds, folivores have a more complex digestive system.

Let's break this down a little further by great ape species using rough numbers from a variety of accessible reference sources, for example, *Primates of the World* by Jean-Jacques Petter and François Desbordes (2013). While great ape eatables are related to those consumed by humans, Chapter 4 will cover the evolutionarily salient human diet and dentition in more detail.

Chimpanzees and Bonobos

Chimpanzees can spend half the day eating with even more time allocated for food resourcing. There are about 180 types of vegetation covering 140 tree and plant species. Commonly consumed are 155 or so plant types consisting of fruits (50 percent), leaves (about 25 percent), buds (about 25 percent), supplemented by seeds, flowers, stalks, inner plant tissue, along with tree bark and resin. In all, about 230 different plant foods are eaten. Chimpanzees are known to self-medicate by eating plants that act like antibiotics, fight internal parasites, and work against malaria (Kaplan & Rogers 2000). Other great apes similarly medicate themselves (Strier 2017). Insects, bird eggs, birds, and small mammals are also eaten on occasion. Soil is consumed for its mineral content. When available, fruits are the largest part of the chimp diet while leaves are eaten often throughout the year, followed by seeds and then insects like termites and caterpillars when available.

Famed primatologist Jane Goodall (1986) conducted extensive field research in the Gombe Stream area of Tanzania and says that chimpanzees "crave variety" in their diets with up to a dozen different foods on any given day. Covering two years of data observations of male and female chimpanzees, the vast majority of foods eaten (65–95 percent) consist of fruits and leaves. Insects and meat range from less than 5 percent up to 20 percent any year, with some months near zero. Even so, on any given day, chimps might eat insects, but certainly not prized meat. Insect and meat eating vary widely month to month, and these percentages, gathered across 1978–1979, are only approximations. During leaf eating, insects might have been eaten accidentally, which developed the taste for a novel food. Likewise, over time as chimps resourced eggs as a food, they might have discovered that they could eat fledglings. Though typically not with overt aggression, chimpanzees will compete for food and food sites, yet that occurs between different females on behalf of their young where there is, simply, displacement of someone else from a feeding site.

While Goodall admits that meat consumption is not unusual, a table she provides indicates that from 1960–1981 there were only 221 red colobus monkeys eaten at Gombe. That would be less than one monkey consumed for any group during a time span of over twenty years. Worth noting is that a chimp group could consist of over, but often less than, 100 individuals. On average, an adult red colobus weighs up to 20 pounds, where the chimpanzees would typically "hunt" juveniles. Gisela Kaplan and Lesley Rodgers (2000) note that in another ten-year study, chimpanzees at Gombe killed and ate nearly 400 colobus monkeys, fewer than one per day multiplied by any number of individuals in a group across the community, depending on what was shared with whom. David Watts and John Mitani (2015) indicate that this hunting in Kibale National Park, Uganda, has increased over time since the red colobus is easy prey. Between 1995 and 2014, 912 colobus monkeys were target kills. The population would not decline to near zero numbers, since fewer victims would be harder to get. Other meat is available, and there does seem to be a significant increase there. Importantly, Watts and Mitani point out that chimpanzees are not "obligate carnivores" like big cats. If the monkeys are available and easy prey, since they share a similar habitat, the encounter is advantageous for the chimp.

Male chimpanzees dominate hunting of game and consume most of the meat of vertebrate animals up to about 22 pounds, even in distribution (Wood & Gilby 2017). On the other hand, Peter Andrews (2020) ranks the killing and eating of meat by chimpanzees low on the scale; they eat more insects. Meat eating among chimpanzees can be prompted by, among other factors, a dry season when staple plant foods are scarce (Pobiner 2020). Diminished ranging because of human encroachment might account for some of these increased monkey kill rates over time. Nevertheless, this hunting is not easy, usually performed by a group, and must be learned. Were the hunting to subside, it's possible younger chimps would hunt less often, if at all. As I've been suggesting throughout, humans could apply that type of cultural learning modification to the cessation of meat and dairy eating.

Getting back to what Goodall reports (1986) in the older study, other animals were less frequently eaten, though over that same time span in the 1970s, a total of 194. In all, 415 various prey were eaten, including some chimps, over a twenty-year period. That's meat at nearly 60 percent for some chimpanzees who likely shared with others or used the meat, frequently, as a sexual incentive or a status alliance commodity. It's important to stress this because the meat eating carries a vital social function (Strier 2017). Keep in mind that this 60 percent covers a span of two decades and does not represent a daily, weekly, or monthly total. Organisms require food to survive, but evident in the meat eating of some chimps in certain locations at various times foodstuffs can be cultural, the message of this book. Shirley Strum (1987) observed olive baboons not far from Goodall's Tanzania and noticed changes in a troop based on the hunting skill of an individual who, eventually, enabled other males to help or copy him. Strum noted 100 kills in 1973 of baby gazelles, ten times the amount of monkey kills by chimps at Gombe. This hunting was not universal baboon behavior, Strum says, since baboons are not true carnivores. The hunting over scavenging seems cultural, evident in how it tapered off among these baboons to very little in time. Hunting and meat eating, past and present, are human behaviors shared with some primates, though in wild populations of chimps meat consumption varies because of a number of factors, including differing cultural practices. Vicky Oelze, et al. (2020) used hair isotope data over seven months for two groups of western chimpanzees and found variations of meat consumption due to microhabitats, sex, and social groups.

Yet it's not my intent to cherry-pick data, and in fact, one can see the difference between the study by Goodall (1986) and the one by Watts and Mitani (2015). The truth is that chimps certainly like meat and will hunt for it, even if opportunistically. In one case (Nakamura, et al. 2019), wild chimpanzees confronted and deprived a leopard of the kill. This behavior is not surprising given their very close genetic, morphological, physiological, and social behaviors to humans who have literally eaten some animal species to extinction (Sale 2006). At the same time, to be fair, Peter Andrews and R.J. Johnson (2019) say that since the animal matter consumed by chimpanzees is less than 10 percent, they are not even to be considered omnivores. Ape digestive tracts evolved to eat leaves and fruit, not animals. Ape and human digestive anatomy is, however, overall similar. The principal difference is that the human digestive tract is shorter, since early *Homo* began eating more cooked meat that was easier to break down than leaves.

Perhaps, like humans, chimpanzees fortuitously acquired a taste for meat as an unusual food source in selection pressures. A key difference, of course, is that chimpanzees, like other great apes, have not dominated their continents. Their low birth rates and evolved population limits are unlike, to borrow from Mark Moffett (2020), the "human swarm." While "parks" are large, chimpanzees are now restricted to zones smaller than they would have had historically before the human incursion. In this case, encounters between chimps and monkey prey might be on an increase because of proximity and feeding stress for all of the many inhabitants in these circumscribed areas. Already in an earlier time, George Schaller (1964) remarked how improperly managed national parks hem animals into a confined area with false borders forcing them to compete for natural resources that, in turn, can lead to environmental problems.

Two bonobo groups tracked and followed for a five-year study (Samuni, et al. 2020) show cultural differences in favored meats in spite of similar habitats and resources. Such prey "hunting" produced only thirty-one anomalures (a small rodent), for one group, and eleven duikers (a small antelope), for the other group. This study validates that some food behavior exists independent of ecology and is, rather, based on cultural preference. Contemporary humans, too, can alter their food choices and avoid processed meat and dairy for health and environmental reasons. We might want to consider to what degree humans need to be predators of any form of animals for nourishment. For humans, food is ecology as well as culture. Hence, we require a restructuring of our food preferences to boost the sustainability of ecosystems, not pressure them into deleterious stress.

Regardless, bonobos do not regularly hunt small mammals, perhaps because they tend to be less aggressive than chimpanzees and have a more plentiful supply of other foods (Strier 2017). Like chimps, they will eat insects and eggs, even small fish, but thrive mostly on fruits, leaves, flowers, stems, and roots, feeding on about 145 different plant foods. Comparable to chimpanzees, bonobos are remarkably close to humans in terms of genetics, morphology, physiology, and social behaviors. However, it is worth noting that male and female chimpanzees and bonobos in the wild carry very little excess body fat (Newson & Richerson 2021) and can survive as well as human hunter-gatherers (Wood, et al. 2017a). Obviously, in contrast, many modern humans in developed, industrial societies are subsisting on the wrong diet.

Gorillas

Gorilla digestive systems are not typical for folivores. With all great apes, digestion is oriented toward frugivory with gut similarity between gorillas and chimpanzees, says Juichi Yamagiwa (2004). Gorillas and chimpanzees are close genetically and similar in their morphology and cranial-dental aspects but different in how they socially organize themselves ecologically, say Juichi Yamagiwa, et al. (1996) in a study of feeding overlap in the Kahuzi-Biega National Park, Zaire. Mainly as folivores, gorillas fuse around one male with several females in leaf-rich locations and feed among about 230 different plant foods. Chimpanzees, primarily as frugivores and insectivores, engage in fission–fusion groups centered on male associations, spreading out for feeding during the day. Gorilla and chimpanzee ranges and diets can overlap. For example, lowland gorillas eat fruit and insects and, like chimps, will nest in trees, creating an interspecies competition, mainly during food scarcity. Among lowland gorillas, those in the higher eastern areas can show dietary similarity to mountain gorillas. Eastern lowland gorillas with small and scattered populations, who range in nether regions, feed more on fruits and insects. Gorillas seem to range more widely than chimpanzees, utilizing a variety of vegetation to feed and nest. In the lowlands there's much fruiting food overlap for gorillas and chimpanzees, but chimps consume more parts of a plant absent fruit in the higher elevations.

Looking particularly at mountain gorillas, Dian Fossey (1983) tells us that favored foods include ferns, buds, and the wood covering, flowers, and pulp of the *Veronia* tree. Gorillas also feed on fruit and berry shrubs and giant lobelia. Before Fossey, Schaller (1964) also spoke about the vegetarian diet of the mountain gorilla. There are about 55 plant species across several zones in the mountains of eastern Africa. Gorillas can find there for their delectation thistles, nettles, celery, and *Galium* vines. There's also bracket fungus, a tree parasite like a mushroom, that is scarce and a favored food. Dirt is consumed for potassium and calcium, as well as bark and roots. Leaves, shoots, and stems make up about 85 percent of the mountain gorilla's diet and fruits only about 2 percent. There is not much food competition. In addition to leaves, grubs, worms, and snails are also eaten, but considerably less in comparison to vegetation (Jaffe 2019).

For gorillas, some areas have a greater number of foods than others. For example, the Virunga montane has approximately 80 foods (with about 40 plant species) whereas the Lopé Reserve has about 210 foods (with about 160 plant species), according to Yamagiwa, et al. (1996). This differential range demonstrates gorilla dietary flexibility. The Yamagiwa study shows that in 256 fecal sample analyses of 54 gorillas and 394 fecal sample analyses of 22 chimpanzees, there was less than 1 percent insect remains for the gorillas to 30 percent for the chimps, with 2 percent showing mammal remains for the chimpanzees. Thus, their foods differed dramatically in spite of their genetic and habitat similarities. Chimpanzees can eat some meat, but much depends on the community, location, and other factors already mentioned.

Encounters between these two species in fruiting trees were non-aggressive, say Yamagiwa, et al. (1996). At the same time, meetings at these trees between conspecific gorilla or chimpanzee groups revealed aggressive posturing and vocalizations. So-called interference competition is far less between different species than among groups of the same species across several study sites. This conclusion, says Yamagiwa, is based on foraging patterns and what exactly is eaten from the fruit tree. Gorillas might not necessarily eat the actual fig, preferred by the chimpanzees. These patterns of low aggressive behavior are evolved adaptations, opportunistic fruit eating by gorillas. As mostly folivores, gorillas can range into higher areas where there is less fruit, favored by the chimps. Most likely, here is a diet stemming from ecological pressures because of the colder and drier climate beginning in the Pleistocene.

The point, certainly, is that ape biological and cultural adaptability must now be mimicked in human apes in a modification toward a vegan diet as we compensate for poor health and correct climate change.

Orangutans

Orangutans, notes renowned primatologist Biruté Galdikas (1999), rely on a multifaceted compound diet of fruits, nuts, leaves, bark, sap, shoots, stems, honey, fungi and other such foods, including insects. Orangutans closely inspect and eat up to about 400 different foods, mostly plants. Gisela Kaplan and Lesley Rogers (2000) indicate that because of their need to travel and recall fruit locations, orangutans have excellent spatial and temporal awareness. Likewise, Anne Russon (2004) says that orangutans demonstrate a high degree of intelligence in how they identify, handle, and manipulate foods for sustenance, one reason why she dubs them as "wizards of the forest." Their staple nourishment comes from wild, ripe fruit like durians mangosteens, mangoes, merang, belale, jackfruit, snakefruit, ramutans, and banitan nuts. There could be nearly two dozen fruiting trees and vines monitored, each ripening at different times of the year across various locations of the Asian forest.

Though rare, orangutans have been seen to eat small mammals, but this could be driven by food stress or the physical requirements placed on a lactating female. The individual home range of an orangutan will vary depending on the location (e.g., Borneo or Sumatra). Estimates suggest that for a Bornean orangutan, a range could be about several hundred square acres, more for a Sumatran orangutan. This estimate depends on whether the orangutan is a resident, a commuter, or a wanderer, say Junaidi Payne and Cede Prudente (2008). Adult males tend to roam more widely, perhaps in search of mates. Ranging is important in terms of sourcing preferred fruits high in carbohydrates and proteins, though orangutans will eat bitter fruit, less nutritional figs that are abundant in many species, and leaves. A fallback, famine food is a tree's exterior covering.

Captive Ape Diet

As part of this project, in 2020 I contacted twenty-one primate sanctuaries (Rosenman 2020, pp. 337–338), including one research facility, across North America and Africa to understand captive ape diets. Great apes are in sanctuaries for a number of reasons. They might have been used in biomedical experiments for surgeries, drug trials, and other treatments. They might have been in road shows or the entertainment industry. They might have, as infants, witnessed their family get slaughtered for bush meat so they could be trafficked as illegal pets. Their plight represents some of the most callous human attitudes toward "animals" and their habitats for commercial gain. Meanwhile, there are many caring people and groups helping our ape cousins, evidenced by the rescuers, sanctuaries, and financial donors.

There was a robust 40 percent response rate providing a wealth of information summarized here. Although one could infer the answers, these were the questions asked.

1.   What are some of the staple foods offered to apes?
2.   Are apes allowed to forage indoors/outdoors and how?
3.   Do apes eat any type of animal products provided by the sanctuary?
4.   Has anyone observed an ape eating small animals or insects in the "territory"?

Depending on the geographic location and time of year, the diet for large, captive primates is primarily a mix of fresh vegetables, fruits, nuts, etc., ranging from onions, tomatoes, carrots, potatoes, sweet potatoes, peppers, broccoli, celery, cauliflower, avocado, cassava, eggplant, maize, green beans, potato greens, corn, legumes, cabbage, yams, rambutan, cucumber, watermelon, jackfruit, pawpaw, assorted melons, lemons, limes, assorted berries, mangoes, apples, oranges, grapefruit, pomelo, grapes, pears, pineapples, avocados, bananas, rice, peanuts, peanut butter, sunflower seeds, assorted unsalted nuts, posho, millet or soya flour, and other various beans. There's often an abundance of leafy greens like kale, collards, dandelion, chicory, spinach, and different kinds of lettuce. One North American sanctuary said that it leans more to the leafy greens than commercially produced fruit out of concern for sugar content. Vitamin and mineral supplements might be administered separately or through fortified, measured primate chow. If the chow is not vegan certified, the calcium it contains is possibly from dairy and it might include taurine from animal bile. Omega 3 and linoleic acids can be found in flax seed or corn oils, which are used in cooking, if there is any.

Some dairy is provided, like cheese, cottage cheese, or yogurt, and there might be a routine hard-boiled egg. Not surprisingly, motherless babies receive milk and Cerelac, a nutritious instant cereal that's easily digestible. In some cases adults are offered milk. New rescues from the wild have other foods like passion fruit, papaya, and tree tomatoes. Sometimes the newly rescued apes are given eggs, though apparently these are not to everyone's liking. The freedom to make individual choices from a variety of plant foods is important and relevant to the argument here about a human shift to veganism. From their enclosure lands, foliage from a diversity of trees, vines, and grasses are consumed. Correspondingly, human vegans can enjoy a wide assortment of delicious vegetable, grain, nut, or seed foods, whether raw or cooked.

Sarah Huskisson, et al. (2021) tested captive gorilla, chimpanzee, and Japanese macaque species in two food-preference tests where the primates seem to select foods even by chance (i.e., not knowing an outcome) with a likelihood of not getting the preferred food. Availability and diverse conditions affect primate food choices in the wild. A lot focuses on risk aversion relative to chance-taking for a higher payoff, evident even among humans. Huskisson concludes that the primates might not necessarily have been taking a chance in the tests. Rather, their uncertain choice might be strategic to maximize reward anticipating that, regardless of choice, some type of food would be offered. For many people in wealthy, industrialized nations, nutritious and healthy vegan food options are plentiful, so there is little danger of not finding something palatable.

Sanctuaries try to mimic foraging by cutting up a mix of fruits and vegetables and spreading them out in a large yard across acreage. In North America during winter, indoor foraging is provided. During an outdoor browse, apes naturally have been observed eating an insect or one they have pulled off a social partner. In one case, there's a sanctuary with nearly 100 acres of forest, and the apes are semi-captive. In the African woods they forage on figs, *Aframomum*, and the leaves and fruits of *Canarium schwinfurthii*, a large, native tree. In another African sanctuary some chimps feed on leaves and bark. Leaves are not only food but are used, of course, to make day resting or night sleeping nests. The more time they spend outside, according to the research study facility with chimpanzees, the more likely they are to snack on insects. A North American sanctuary says it has not observed any of the chimps eating insects outside.

A few African sanctuaries say no chimps outside have been observed eating small mammals, but predictably they do fish for termites, ants, or lake flies. Two North American sanctuaries experimented with feeding chimps mealworms, but there was little interest. A sanctuary in North America says one chimp has been observed on two occasions eating frogs, and several have been seen eating wasp larvae. Naturally they also like honey. Another North American sanctuary indicated that on their large islands one chimp in particular has been known to eat turtles. Since many sanctuary apes have lived mostly in captivity, they might not be attuned to small mammals. In parts of Africa, chimps in the wild have been seen eating tortoises (Pika, et al. 2019). It's reported that some chimpanzees have eaten lizards. Another sanctuary witnessed a chimp kill but not eat a lizard. One sanctuary survey respondent said, in answer to question 4 about meat eating, that chimp food tastes develop culturally within a family, group, or community. This statement aligns with much of what's argued here, since food and diets are cultural. Humans need to embrace the ecology of vegan sustenance for health and environmental reasons. An ape in one sanctuary who employs American Sign Language has requested meat, perhaps because she had it previously in captivity.

Notwithstanding meat eating anomalies in this chapter, granted some severe, plant foods and plant-based proteins are a motif. The immense variety of fruit and vegetation show how feeding and foraging are important parts of an ape's personal and social enrichment. One North American sanctuary sent a comprehensive list of fruits, vegetables, and greens that totals 119 items, not including herbs. With some qualification as noted, it's fair to say fruits and plants are dominant, mainstay foods among great apes in the wild and in captivity, were a chief food supply for our australopith ancestors as explained in the next chapter, and predictably are core nourishment for contemporary vegans, whether raw or cooked. Understandably, the effort is for a sanctuary to imitate as much as possible the natural diet of an ape in the wild, and in that case there is basically no meat with very little dairy, justifiably for nursing infants, to supplement the protein derived from nuts and leaves. At the same time, the diet can include commercially produced "artificial" substances, like nutrient infused primate or monkey chow. In answer to the third question above, one African sanctuary flatly stated that not any animal products are given to the apes.

Just to be clear, the point in this chapter is to demonstrate that, on the whole, animal flesh is not an essential or staple food of great apes or other primates. Researchers Lesley Newson and Peter Richerson (2021) go as far as saying that today's chimpanzees probably don't differ much in terms of morphology, diets, sociality, and life history from our ancestors. Let's take this factual information and, in the next section, expand on it in a theoretical manner in line with my overall argument. While food is cultural, foodstuff farming and production might not be ecological. Animal products are not essential to all human diets, especially in how we compromise our health and the environment in farming and eating so much of it.

Animals as Food

In his book *Thinking Plant Animal Human* (2020), philosopher David Wood contemplates where humans belong in nature, part of my theme. Our ethics are reflected in the systematic deforestation of biodiversity and the eating of billions of animals each year. As much as we hurt the natural world, there's a ricochet effect harming us. The word "animal," says Wood, is a deliberately vague appellation giving humans license to perpetrate violent acts against other species. That's true, and Wood rightly notes how the word animal is a human injustice to many individuals of many species, a violent signature of otherness against them without ascribing a name. We're proud to promote "humanity," but humanism is suspect with its typically bigoted self-interest. If, as some believe, we are supposedly exceptional, it's difficult to explain why we've established economies designed to fail, engineered pandemics because of eating wild meat, and become architects of a blighted and doomed environment.

Joseph Henrich (2016) says, with debatable emphasis on human cultural cooperation, that we have "ecological dominance," as if that's good, and "ecological success," which is exaggerated. Calling modern humans ecosystem engineers might not even be accurate. Instead, we tend to transform and control ecosystems in negative ways contrary to the needs of animal inhabitants to satisfy our commercial or consumerist desires. This landscape "management" might have begun with African hunter-gatherers in the Middle Stone Age, circa 100kya, on a small, practical scale opening forests or grasslands to exterminate pests or flush out prey (Thompson, et al. 2021). Confined land impacts like this are not comparable to the massive, technological exploitation of biodiversity or machine rampaging deforestation by most modern humans who mainly look at the outside world and consider how it will advantageously serve them.

According to Wood (2020), many of these evils stem from the Western, corporate ethic of privilege. We see it in other societies as well, and orchestrating a movement away from destructive attitudes toward healthful veganism is a righteous purpose. There are many individuals and groups fighting for animals, the environment, and marginalized people. Overall, the question focuses on how, in spite of Enlightenment rights and rationality, we've willfully eroded our sense of responsibility for anything not human and even for some humans. We deny our own vulnerability and evolutionary connection to other species, what primatologist Frans de Waal (1999) has called anthropodenial. This renunciation often erupts in the sometimes wild reactions to perceived threats from other "races," genders, or nations. We've lost any sense of our cognitive ecology that viscerally connects us to forests and other animals. In terms of eco-psychology (Barrett 2011), there's no division between an organism and its environment, but we've "evolved" to erase any natural connections and to erect barriers between ourselves and ecosystems. With animal farming, humans have dangerously separated their minds from the environment as if there are no consequences for that breach.

Through hunting and a tribal mentality meat eating is masculine. Not to alter his meaning too much, but when Wood (2020) talks about "animal ethics" he signals the human obligation toward nonhuman species. There's a twist here since one can see animal ethics as the biology of morality or how our moral systems are rooted in our deep evolutionary past and connection to mammals. To say we humans have a principled duty to animals simply puts us at the top of a false hierarchy. If you've ever seen a gorilla give birth and her immediate reactions of gentle cradling, intense eye contact, hugging, and kissing, how could humans alone believe no other living beings have moral stock in their ethical existence. On the contrary, wildlife and farm animals have withered to objects in our collective consciousness away from the self-contained agents they really are. Animals are not against us in a binary opposition but perform actual work caring for and maintaining the biosphere, whether bees, worms, coral, primates, beavers, birds, or whales.

Wood (2020) talks about people in the Asian Gobi Desert who use goats to graze sparse vegetation not fit for human consumption. The goats provide milk and, in turn, their bodies as meat. As is now well known, the carnivorous diet of Inuit people scattered across Siberia, Alaska, Greenland, and Canada has proved unhealthy. They age quickly with an early onset of heart disease and bone loss. This is only a survival diet. Consumption of too much animal protein is not optimal nutrition, so human diets need to be balanced in favor of the starches, oils, and protein from plant foods, as has been demonstrated in an analysis of hunter-gatherer diets (Cordain, et al. 2000). Ideal nourishment can be achieved in a vegan culture with the added benefit of reducing global warming and animal cruelty.

The question is why people live in such a harsh environment; the answer is that it's their tradition. One wonders if this has been a wise coevolution. This situation is even more complex than what Wood intimates since the Gobi steppe ecosystem is expansive grassland that sustains biodiversity across Mongolia. Nevertheless, changes in governmental policies there have allowed citizens to graze any animal for profit. The result is that herders are depleting these pasturelands. One suggestion is for international trade laws to alleviate such untenable land use (Eitel 2020) when economic policies erode environmental viability. On the other hand, modern, industrial societies that farm and eat animals have not fared much better. Clearly, some cultures depend on animal meat as a prime resource, but change is possible. The aim is to shift attitudes in consumerist cultures toward the healthy, nutritionally rich vegan options available. That's a start and could include communities in many modernized, industrialized cities.

Most people can have anything shipped from anywhere, certainly even meatless food. As for the Gobi dwellers, their position of owning another species as a source of nutrition or economic gain can be morally problematic for some thoughtful people. Except, perhaps, for the Gobi Desert dwellers and others similarly situated, many well-fed urban and suburban residents can reform food traditions through any of the many vegan options readily available or accessible by delivery. A culture does not need to eject its heritage. Rather, there are viable ways to shape food into something healthier, environmentally friendly, and cruelty free. Humans, without doubt, are capable of making a plant-based dietary upgrade. Besides, whatever one's ethnic orientation, in a modern urban setting she knows virtually nothing detailed about how her breakfast dairy, lunch meat, or fleshy dinner has been produced or processed.

In fact, the species mentality is part of the problem. The genetic similarity between pigs and humans is about 85 percent, but we eat them in a form of cannibalism. Perhaps that's an outrageous claim, but one could project that idea into a meat-centric society. Heart valves and skin grafts from pigs are harvested for surgical use on humans. So humans literally and figuratively live off pig flesh. How we understand ourselves is lined up with our contemplation, or not, of animals amid the violence of carnivory. We somehow disengage (animals as pets) but simultaneously endorse the genocide of the meat and dairy industries. Humans have generated a cult of animal meat. Human history is marked and marred by our mostly rapacious attitudes toward animals, as well as to people and groups we see as outsiders.

The human attitude is one of dominance over other species, even though we too are "animals." We don't see that superior positioning in, for example, how great apes live peaceably among others in their forest homes. Yes, there are intermittent exceptions, as noted, among chimpanzees. If we say we value life, that all lives matter, and if protein and other nutrients are easily obtained from plant foods that can be cooked and seasoned to taste, there's no rational reason to eat meat. Wood (2020) suggests it's our need to exert sovereign control over animals. All biological organisms seek to control their environment (see Chapter 2), but humans seem to have socially constructed this basic need into a destructive desire. There's a cross-cultural taste for ascendancy. We can, however, adapt new desires. Drawing from philosopher Jacques Derrida's notion of *animôt*, or an animal merely as a word and not as the subject of a life, human language is a fierce appropriation of other creatures where they become merely a piece of "meat," says Wood. He cites these statistics as an alarming example. In the United States, during 2017, $69 billion was spent on pets while some 500,000 animals were killed each hour for human food. From a social perspective, there is no interspecies democratic representation for animals, but forms exist in animal sanctuaries.

Ape/Human Culture

Robin Dunbar (2014) says ape culture "pales into insignificance" compared to human culture. That debate is beyond the scope of this writing but relevant to some extent. In fact, work by field researchers reveals a diverse chimpanzee cumulative, community culture related to feeding (Strier 2017; Boesch, et al. 2020). Most modern human societies are insignificant in their complete lack of eco-psychology with no true connection to what's loosely called nature (Tague 2020). Somewhere, evolutionary biologist Richard Dawkins tells a story like this. He looks out his window and sees the birds, happily flying and singing. That's the romantic, human perspective. In reality, Dawkins would say, the birds endure a harsh certainty of environmental stress and such a lack of food that some literally starve to death. In the meantime, most humans have a plethora of stored food but don't always share it with those in poverty.

Dunbar (2014) actually praises modern humans for their ability to "detach" from reality via story, but their estrangement from the natural world, as illustrated in the Dawkins example, is the unaddressed problem from the eighteenth century onward. Succeeding the work of Robert Boyle and John Ray in the late seventeenth century praising the adaptive designs in nature by a divine hand, eighteenth-century thinkers surrounded by efficient British technology, commerce, and machines truly believed that godly nature was set before them to be exploited (Radick 2003). This reformist attitude tried to promote the status quo by tying together science and religion. If humans were globally cooperative and not typically at war or in conflict, had not ravaged other populations and nature, were open to true reform regarding poor people, hungry people, minorities and under-represented groups, women, children, etc., then one might buy into the allegedly high rank of *Homo sapiens*. Primarily because of our detachment from and lack of empathy for the inhabitants of nature outside of our narrow spheres, it's difficult to see humans as exceptional. The truth is we have plundered the earth and each other's populations to such an extent that we've faced nuclear obliteration, pandemic mass death, and many corporate mistakes of horrible proportions, like oil spills, gas leak explosions, and nuclear power plant contaminations.

Without expounding details here, philosopher of science Chris Herzfeld (2016, 2017) shows how confined apes from the nineteenth to twentieth centuries in zoos and researcher homes ably appropriated the manners, customs, diets, language symbolism, and objects of their captors in an effort of *savoir-faire* to adopt human culture. Herzfeld cites many instances of this almost seamless transition, a blurring of the boundaries between great and human apes. This point is noteworthy because modern humans, in a similar mode of adaptation, have the innate know-how for assimilating into vegan culture since it's in our DNA. The assertion that meat is essential food for humans is patently untrue, since great apes get plenty of protein, nutrients, and vitamins from fruits, leaves, nuts, seeds, etc. Many other people or less "developed" societies are plant-based for economic or religious reasons and have not perished. This chapter's theme about how our fallacious dependency, as apes, on meat leads us into the next chapter, where the vegetarian diet of the earliest humans, australopiths, will be considered.

References

Andrews, Peter. 1985. "Species Diversity and Diet in Monkeys and Apes During the Miocene." *Primate Evolution and Human Origins*. Russell L. Ciochon and John G. Fleagle, eds. Menlo Park, CA: Benjamin/Cummings Publishing. 194–204.

Andrews, Peter. 2020. "Last Common Ancestor of Apes and Humans: Morphology and Environment." *Folia Primatologica* 91: 122–148. doi:10.1159/000501557.

Andrews, Peter and R. J. Johnson. 2019. "Evolutionary Basis for The Human Diet: Consequences for Human Health." *Journal of Internal Medicine* 287 (3): 226–237. doi:10.1111/joim.13011.

Barrett, Louise. 2011. *Beyond the Brain: How Body and Environment Shape Animal and Human Minds*. Princeton: PUP.

Berthaume, Michael A., et al. 2020. "The Landscape of Tooth Shape: Over 20 Years of Dental Topography in Primates." *Evolutionary Anthropology* 29: 245–262. doi:10.1002/evan.21856.

Boesch, Christopher, et al. 2020. "Chimpanzee Ethnography Reveals Unexpected Cultural Diversity." *Nature Human Behavior*. https://doi.org/10.1038/s41562-020-0890-1.

Boonpor, Jirapitcha, et al. 2021. "Heath-related Biomarkers of Profile Vegetarians and Meat-eaters: A Cross-sectional Analysis of the UK Biobank Study." University of Glasgow Poster EP3–33. European Association for the Study of Obesity.

Boyd, Robert and Joan B. Silk. 2017 *How Humans Evolved*. Eighth edition. NY: W.W. Norton.

Clark, Mary Ann, et al. 2018. *Biology 2e*. Houston, TX: OpenStax/Rice University.

Cordain, Loren, et al. 2000. "Plant–animal Subsistence Rates and Macronutrient Energy Estimations in Worldwide Hunter-gatherer Diets." *American Journal of Clinical Nutrition* 71: 682–692.

De Waal, Frans B.M. 1999. "Anthropomorphism and Anthropodenial: Consistency in our Thinking about Humans and Other Animals." *Philosophical Topics* 27 (1): 255–280.

Dunbar, Robin. 2014. *Human Evolution.* London: Pelican.

Eitel, Michael R. 2020. "Adding Tools to the Conservation Toolbox: Can International Trade Policies That Undertax Mongolian Cashmere Provide Relief to Mongolia's Overtaxed Grasslands?" *Animal and Natural Resources Law Review* 16: 41–79.

Fleagle, John G. 1988. *Primate Adaptation and Evolution.* San Diego: Academic P.

Fossey, Dian. 1983. *Gorillas in the Mist.* NY: Mariner.

Galdikas, Biruté M.F. and Nancy Briggs. 1999. *Orangutan Odyssey.* NY: Harry N. Abrams.

Goodall, Jane. 1986. *The Chimpanzees of Gombe: Patterns of Behavior.* Cambridge, MA: Harvard UP.

Hales, Craig M., et al. 2020. "Prevalence of Obesity and Severe Obesity Among Adults: United States 2017–2018." *NCHS Data Brief* number 360. Hyattsville, MD: National Center for Health Statistics.

Henrich, Joseph. 2016. *The Secret of Our Success: How Culture is Driving Human Evolution, Domesticating Our Species, and Making Us Smarter.* Princeton: Princeton UP.

Herzfeld, Chris. 2016. *Wattana: An Orangutan in Paris.* Oliver Y. Martin and Robert D. Martin, trans. U. Chicago P.

Herzfeld, Chris. 2017. *The Great Apes: A Short History.* Kevin Frey, trans. New Haven: Yale UP.

Hicks, Thurston, et al. 2020. "The Relationships Between Tool Use and Prey Availability in Chimpanzees (*Pan troglodytes schweinfurthii*) of Northern Democratic Republic of Congo." *International Journal of Primatology* 41: 936–959. https://doi.org/10.1007/s10764-020-00149-4.

Huskisson, Sarah M., et al. 2021. "Primates' Food Preferences Predict Their Food Choices Even Under Certain Conditions." *Animal Behavior and Cognition* 8 (1): 69–96. https://doi.org/10.26451/abc.08.01.06.2021.

Jaffe, Karin Enstam. 2019. "Primate Behavior and Ecology." *Explorations: An Open Invitation to Biological Anthropology.* Beth Shook, et al., eds. Arlington, VA: American Anthropolgica Association. 190–232.

Kaplan, Gisela and Lesley J.Rogers. 2000. *The Orangutans: Their Evolution, Behavior, and Future.* Cambridge, MA: Perseus Publishing.

Klein, Richard G. 2009. *The Human Career: Human Biological and Cultural Origins.* Third edition. Chicago: U Chicago P.

Lee, Richard B. and Irven DeVore. 1968. *Man the Hunter.* Chicago: Aldine Publishing.

Moffett, Mark W. 2019. *The Human Swarm: How Our Societies Arise, Thrive, and Fall.* NY: Basic Books.

Nakamura, Michio, et al. 2019. "Wild Chimpanzees Deprived a Leopard of Its Kill: Implications for the Origin of Confrontational Scavenging." *Journal of Human Evolution* 131: 129–138. doi:10.1016/jjhevol.2019.03.011.

Newson, Lesley and Peter Richerson. 2021. *A Story of Us: A New Look at Human Evolution.* Oxford: OUP.

Oates, John F. 1987. "Food Distribution and Foraging Behavior." *Primate Societies.* Barbara B. Smuts, Dorothy L. Cheney, Robert M. Seyfarth, Richard W. Wrangham, and Thomas T. Struhsaker, eds. Chicago: U Chicago P. 197–209.

Oelze, Vicky M., et al. 2020. "How Isotopic Signatures Relate to Meat Consumption in Wild Chimpanzees: A Critical Reference Study from Taï National Par, Côte d'Ivoire." *Journal of Human Evolution* 146: 102817. https://doi.org/10.1016/j.jhevol.2020.102817.

Parker, Sue T. 1999. "The Life History and Development of Great Apes in Comparative Perspective." *The Mentalities of Gorillas and Orangutans.* Sue Taylor Parker, Robert W. Mitchell, and H. Lyn Miles, eds. Cambridge: Cambridge UP. 43–69.

Payne, Junaidi and Cede Prudente. 2008. *Orangutans.* Cambridge, MA: MIT P.

Petter, Jean-Jacques and François Desbordes. 2013. *Primates of the World.* Robert Martin, trans. Princeton: Princeton UP.

Pika, Simone, et al. 2019. "Wild Chimpanzees (Pan troglodytes troglodytes) Exploit Tortoises (Kinixys erosa) via Percussive Technology." *Scientific Reports* 9 (7661). https://doi.org/10.1038/s41598-019-43301-8.

Pobiner, Briana L. 2020. "The Zooarchaeology and Paleoecology of Early Hominin Scavenging." *Evolutionary Anthropology.* doi:10.1002/evan.21824.

Quinn, Rhonda L., et al. 2021. "Influences of Dietary Niche Expansion and Pliocene Environmental Changes on the Origins of Stone Tool Making." *Paleogeography, Paleoclimatology, Paleoecology* 562(15): 110074. https://doi.org/10.1016/j.palaeo.2020.110074.

Radick, Gregory. 2003. "Is the Theory of Natural Selection Independent of Its History?" *The Cambridge Companion to Darwin.* Jonathan Hodge and Gregory Radick, eds. Cambridge: CUP. 143–167.

Randolph, Delia Grace, et al. 2020. *Preventing the Next Pandemic: Zoonotic Diseases and How to Break the Chain of Transmission.* Nairobi, Kenya: United Nations Environment Programme.

Rosenman, Debra, ed. 2020. *The Chimpanzee Chronicles.* Santa Fe, NM: Wild Soul Press.

Rothman, Jessica M., et al. 2014. "Nutritional Contributions of Insects to Primate Diets: Implications for Primate Evolution." *Journal of Human Evolution* 71: 59–69. http://dx.doi.org/10.1016/j.jhevol.2014.02.016.

Russon, Anne E. 2004. *Orangutans: Wizards of the Forest.* Revised edition. Buffalo, NY: Firefly Books.

Sale, Kirkpatrick. 2006. *After Eden: The Evolution of Human Domination.* Durham, NC: Duke UP.

Samuni, Liran, et al. 2020. "Behavioural Diversity of Bonobo Prey Preferences as a Potential Cultural Trait." *eLife* 9: e59191. https://doi.org/10.7554/eLife.59191.

Schaller, George B. 1964. *The Year of the Gorilla.* Chicago: U of Chicago P, 1988.

Selig, Keegan R., et al. 2019. "The Frugivorous Insectivores? Functional Morphological Analysis of Molar Topography for Inferring Diet in Extant Treeshews (Scandentia)." *Journal of Mammalogy* 100 (6): 1901–1917. doi:10.1093/jmammal/gyz151.

Sterelny, Kim. 2013. "Human Behavioral Ecology, Optimality, and Human Action." *Evolution of Mind, Brain, and Culture.* Gary Hatfield and Holly Pittman, eds. Philadelphia: U of Pennsylvania Museum of Archaeology and Anthropology. 303–324.

Strier, Karen B. 2017. *Primate Behavioral Ecology.* Fifth edition. NY: Routledge.

Strum, Shirley C. 1987. *Almost Human: A Journey into the World of Baboons.* NY: Norton.

Tague, Gregory F. 2020. *An Ape Ethic and the Question of Personhood.* Lanham, MD: Lexington Books.

Thompson, Jessica C., et al. 2021. "Early Human Impacts and Ecosystem Reorganization in Southern-central Africa." *Science Advances* 7 (19): eabf9776. doi:10.1126/sciadv.abf9776.

Trivers, Robert. 1972. "Parental Investment and Sexual Selection." *Sexual Selection and the Descent of Man.* Bernard Campbell, ed. Chicago: Aldine. 136–179.

Watts, David P. and John C. Mitani. 2015. "Hunting and Prey Switching by Chimpanzees (*Pan troglodytes schweinfurthii*) at Ngogo." *International Journal of Primatology* 36: 728–748. doi:10.1007/s10764-015-9851-3.

Wilson, Michael Lawrence. 2021. "Insights into Human Evolution from 60 Years of Research on Chimpanzees at Gombe." *Evolutionary Human Sciences* 3: e8. doi:10.1017/ehs.2021.2.

Wood, Brian M. and Ian C. Gilby. 2017. "From *Pan* to Man the Hunter: Hunting and Meat Sharing by Chimpanzees, Humans, and Our Common Ancestor." *Chimpanzees and Human Evolution.* Martin N. Muller, et al., eds. Cambridge, MA: Harvard UP. 339–382.

Wood, Brian M., et al. 2017a. "Favorable Ecological Circumstances Promote Life Expectancy in Chimpanzees Similar to That of Human Hunter-gatherers." *Journal of Human Evolution* 105: 41–56. https://doi.org/10.1016/j.jhevol.2017.01.003.

Wood, David. 2020. *Thinking Plant Animal Human: Encounters With Communities of Difference.* Minneapolis: U Minnesota P.

Yamagiwa, Juichi, et al. 1996. "Dietary and Ranging Overlap in Sympatric Gorillas and Chimpanzees in Kahuzi-Biega National Park, Zaire." *Great Ape Societies.* William C. McGrew, Linda F. Marchant, and Toshida Nishida, eds. Cambridge: Cambridge UP. 82–98.

Yamagiwa, Juichi. 2004. "Diet and Foraging of the Great Apes: Ecological Constraints on Their Social Organizations and Implications for Divergence." *The Evolution of Thought: Evolutionary Origins of Great Ape Intelligence.* Anne E. Russon and David R. Begun, eds. Cambridge: Cambridge UP. 210–233.

4

EARLY HUMANS

Now that we've covered preliminaries about biological evolution and primates, let's make the connection to the early human diet of our ancient ancestors. As with chimpanzees, you will see that humans evolved from frugivores who became opportunistic meat eaters. Briana Pobiner (2020) indicates that meat and fat would not have dominated hominin diets before 2mya but for passive or marginal scavenging. Recall that some apes, especially gorillas, rarely if ever eat meat, and yet they are the strongest of ape people. Dates provided here are only approximations, with the oldest occurrence of a species given. I refer the reader to the simplified timeline at the beginning of this book, too. The dates in this chapter and in the timeline might vary depending on which experts are consulted, so they are meant only as a guide. Additionally, there is still much dispute among paleoanthropologists about classifying fossils (Newson & Richerson 2021). Readers will be treated to a number of species names in Latin through some sections in this chapter. Standard texts of paleoanthropology have been consulted to make the point without citing the reams of scientific papers available. Like the chapter on great apes, this one provides important claims for the overarching argument about how we can incline culturally to the ecology of vegan foods.

The key idea to hold is that *Homo sapiens* evolved from a common ancestor to a chimpanzee and, like other, related species, has a shaggy family tree whose frayed distortions defy any straight linearity of descent. Technically, there is no uniform progression. The many allied hominin species who predate us and prepare for our entry were not obligate meat eaters. In no way did they presage the scale of highly processed and chemically contaminated meat and dairy products eaten today. For health and environmental reasons, we can go back to the simpler human diet that was vegan-like, if we choose. Instead, we have somehow maladapted ourselves to processed, supermarket foods. We can return to a plant-based diet and, one could

DOI: 10.4324/9781003289814-5

argue, that from necessity we should adapt a vegan economy of local farming communities manufacturing and distributing veggie foods.

Overview

Regarding chimpanzee and gorilla similarities to humans, look back to *Sahelanthropus tchadensis* (from Chad, circa 7mya), *Ardipithecus kadabba* (from Ethiopia, circa 5.6mya), and *Orrorin tugenesis* (from Kenya, circa 6mya), where these species have reduced canine size and move between obligate to habitual bipedalism (Su 2013). These species indicate that a likely chimpanzee–human split occurred between a mere 5–7mya. Each of these specimens shows either dental or cranial traits closer to hominins, not apes. Barbara Welker (2017) says *S. tchadensis*, like ardipiths (extinct hominin), had thin molar enamel in comparison with later australopiths (extinct hominin) and fed off fruit, leaves, and shoots like chimpanzees, and perhaps some insects. Another species, *Australopithecus sediba* (circa 2mya), had a diet, according to Welker, that was herbaceous, including shrubs and grasses, not like that of other australopiths or paranthropines. One can't rule out consumption of some animals.

W.P.T. James, et al. (2019) affirm how ecological factors, like a cooling environment of circa 2–3mya, altered the human diet with a shift in some species toward occasional animal protein, fats, and minerals. In a similar way, by about 8mya, cooling temperatures killed many European apes while some retreated back to Africa, which had forests and fruits (Andrews & Johnson 2019). James suggests these natural foraging changes, along with bipedalism and cooperative behavior to avoid predators in the savanna account for increased social behavior and brain expansion. Note, however, there are brain gene transcription differences between chimpanzees and humans that could have arisen from other factors of evolution. Human populations migrated fiercely and evolved various adaptations to differing climates and food sources (Yoshida-Levine 2019).

Bipedalism capacities in *O. tugenesis* (circa 6mya) are apparent in an environment that was cooling and drying with expanding vegetal grasslands and shrinking tree forests. Food scarcity would have forced exploitation of new resources for survival. Early hominins like australopiths had shorter arms and longer legs with a big toe like a modern human (Kerryn Warren, et al. 2019). Occasional bipedalism increased the ability to range for ripe fruit as opposed to a smaller range by an omnivore like a baboon. Bipedalism is energy efficient, increases time budgets, and permits carrying. Many ape species went extinct, unless they adapted to these changes by ground dwelling. The bipedal survivors in grasslands would be australopiths (4.4–1.2mya) and, later, *Homo habilis* (circa 2.5mya). Robin Dunbar (2014) goes as far as calling *H. habilis* "a late transitional australopith." *Ar. kadabba* (circa 5.6mya) was most chimp-like and preferred the forest.

Much of Africa was tropical forest well before 4mya; after that, our ancestral relatives spent less time in trees as foot morphology shows (Newson & Richerson 2021). After bipedalism from a number of pressures, including adaptation to heat, evident prominently in *Homo ergaster* (circa 1.8mya) who has an external nose, dentition changed. By 4.2mya, hominins had already developed, compared to chimps, larger, flatter cheek teeth and smaller canines, even in males (Kerryn Warren, et al. 2019). Enamel thickened. These evolved adaptations were for grinding and crushing foods like nuts, seeds, and fibrous material found close to the ground as opposed to nutriments in trees. This is not to say early hominins simply walked out of the forest. Rather, retained ape-like tendencies were adaptations for feeding, nesting, and protection in trees.

As for human evolution, it's like a mosaic, a constellation, or variegated and simultaneous patterns of shared, derived, divergent, and convergent adaptations. There's a difference between homoplasies (traits shared outside of ancestry) and inherited characteristics. It appears *Australopithecus* and *Homo* reflect parallel evolution (homoplasy) over direct evolutionary relationships pursuant to a cladogram of shared, derived features, i.e., family relations over a phylogeny of time. Depending on the relationships considered, evolutionary associations can be established. For example, look at the cranial capacity, not cheek teeth, linking *Homo* and *Paranthropus robustus* (circa 2mya) and *P. boisei* (circa 2.4mya). Paleoanthropologist Richard Klein (2009) sees *Australopithecus africanus* (circa 3mya) as a "plausible ancestor" to *Homo*, more or less. Others (Welker 2017) might place *Au. sediba* (circa 2mya) more centrally, but splitting hairs like this is outside the boundaries of this argument. From about 4.5–2.3mya covering *Ardipithecus* with conservative molars geared for frugivory to *Au. africanus* (circa 3mya) with molars more specialized for tough foods, say Mark Teaford and Peter Ungar (2000), hominin dietary competencies in teeth size and enamel evolved in response to climate and available resources. Phytolith (mineral plant particle) analyses of *Au. sediba* reveal a diet from both grasslands, similar to that of savanna chimps, and forests, including fruit, leaves, and bark (Henry, et al. 2012; Welker 2017).

Just as some of our direct or related ancestors were highly adaptable regarding their food, so we can be through cultural adaptation. Recent field research (Martin, et al. 2020) implies that microevolution pronounced in the morphology of an early *Paranthropus robustus* (circa 2mya) in one South African location, as opposed to another locale, could have arisen from pressures related to changes in ecology, not social behaviors or sexual dimorphism. These findings support claims of species ontogeny deriving from geography and climate then and now. This does not mean, however, that present-day people cannot change diets culturally. Evolutionary responses to a changing climate and food sources helped *P. robustus*, a contemporary of *Homo erectus* (circa 1.8mya), survive whereas australopiths did not. Today, we cannot rely on rapid biological evolution to armor us against climate change or expect all other species to suddenly adapt dramatically to shifting environments. Because they are so widespread and exert such force on climate, humans can be the instigators of an improved global environment through their own social evolution.

Anyway, based on dental morphology, one of many likely *Homo* ancestors is *Australopithecus garhi* (2.5mya). It's difficult to say whether *Homo habilis* (circa 2.5mya) or *Homo rudolfensis* (circa 2mya) lead to later *Homo*, which seems reserved for *Homo ergaster* (circa 1.8mya), arguably the first of hunter-gatherer appearing before the disappearance of *H. habilis* or *H. rudolfensis*. What seems clear is that these species' lineages have brain enlargement, smaller cheek teeth, habitual bipedalism, and the ability to use stones as tools, all of which foreshadow anatomically modern humans (Paskey & Cisneros 2019). *Australopithecus afarensis* (circa 3.9mya), and particularly the gracile *Au. africanus* (circa 3mya), though living before early stone tools (circa 2.5mya), possessed hands anticipating those of human dexterity for the manufacture and use of implements in later hominins. *Au. afarensis* would have been small at a maximum height of 5 feet and weight of 100 pounds, liable to predation from large hyenas and big cats, explaining its frequency in forest trees. This species with thick enamel, says Welker (2017), ate tough foods like tubers and roots.

Crude tools were used for de-fleshing meat from bones (later, butchering), bone marrow extraction, and possibly as projectiles. Similarly, stones as tools could be used to dig root vegetables, perhaps their first occurrence after wooden kits (Yoshida-Levine 2019). Australopithecines were essential frugivores who ate some meat when available and likely had some primitive stone devices. Since living apes are tool makers and tool users, they probably inherited that capacity from a common ancestor with hominin australopiths who continued in time to live along with *Homo habilis* (circa 2.5mya). Not all ancestral humans had the hand morphology of precision grip as *H. habilis* and later humans. There is controversy, says Richard Klein (2009), about how far back to date the human thumb. Kerryn Warren, et al. (2019) also debate morphological evolution. Dunmore, et al. (2020) report how the *Au. sediba* thumb (circa 2mya) was diversely used for manipulation and tree locomotion. Robust australopiths survived well past 500kya primitive (Oldowan) tools into a period of more advanced technology (Acheulean), but there's no indication of an australopith stone tool industry as there is for *Homo*. Robust australopiths relied on manipulating vegetation by hand, it appears. Less emphasis on a vegetarian diet, a larger brain, less powerful jaws and smaller teeth, seem to account for *Homo* tool adaptation to carnivory even if, as is likely, not through hunting or systematic kills, but by scavenging carcasses.

Early people from 2–1.5mya, at least, were poor hunters. Researchers like Klein (2009) might suggest that other animals, similar to porcupines who gnaw or ungulates who trample, could account for striation found on purported bones used by *Homo* for food sources. Compressed sedimentation could also account for marks on fossil bones attributed as human food. Many site bones (e.g., antelope) suggest other animal carnivory (a lion kill) followed by human scavenging with fortuitous or roughly made tools to de-flesh bone. This is not absolutely clear, says Klein. Chimpanzees prefer raw meat and rarely scavenge. There's a distinction between apes and early humans who mostly fed off carcasses left by others, therefore opening an advantageous ecological trait to build from, i.e., implement manufacture. The use of devices like a randomly sharp stone to de-flesh bone or a rock hammer to smash bone for marrow could, at first, have been accidentally discovered from a digging tool and then repeated and copied in a cultural manner, as is common in chimpanzee communities (Boesch, et al. 2020).

Primate Origins, Australopiths, and the Question of Who's Human

Klein (2009) suggests a primate ancestor in an insectivorous eating Cretaceous animal at over 65mya when continents were still close together. (At 200mya there was one continental mass.) There was a mild global climate with temperate forests in both hemispheres with mostly non-flowering plants in the pre-Cretaceous period. The flowering plants appear early in the Cretaceous time and hence fruits, berries, and nectars, creating an explosion of plant-pollinating insects. With a diversity of plenteous food, birds and mammals diversified. While early primates undoubtedly ate insects, the bluntness of their molars indicates that they also ate fruits, seeds, and other vegetal foods. These primate ancestors of the Cretaceous period might have been more squirrel-like (Clark, et al. 2018). Primate specialization probably arose as movement to fruits at terminal branch sites, where insects would thrive, increased along with enhanced vision.

During the late Cretaceous around 65mya, Palocene Plesiadapiformes evolved new tooth morphology by adding seeds, fruits, and vegetation to their insectivorous diets. Later radiations, like lemur and tarsier types, evolved grasping hands and brain reorganization more for vision, also in terms of widening the diet to include fruits and flowers at branch tips. At the middle Eocene, circa 45mya, the focus was less on insects and more on fruits and leaves. By about 25mya in the late Oligocene, higher primates evolved into forms closer to extant apes and monkeys, where dental morphology put greater emphasis on leaf eating among monkeys and fruit eating among apes. After 17–16mya, or the later Miocene, which is more cool, dry, and seasonal, we see the beginning of modern apes with new dentition for hard foods like seeds, nuts, or bark as fruiting seasons waned.

Early in the Miocene, as reported by Peter Andrews (1985; Clark, et al. 2018), there were more apes than monkeys. Terms such as frugivorous and folivorous should not be applied adamantly, given a primate's range and feeding, seasonality, and fallback foods. Originally, from the early to the middle Miocene, it seems the anthropoid diet was frugivorous, based on fossil analysis of their large incisors and lowly-rounded molar cusps, but leaves and seeds were certainly eaten, says Andrews. In the meantime, some species developed specialization for leaves to expand a seasonal diet and live among frugivorous hominoids. Monkeys can tolerate allelochemicals (a defense agent) more than apes and eat more unripe fruit, a selective advantage to process toxins. Conceivably in competitive response to monkeys, hominoids developed thicker enamel, small anterior teeth, and large posterior molars suggesting an omnivorous diet. Another specialization, according to Andrews and lacking in monkeys, is the hominoid appendix, an immunological device perhaps to neutralize vegetable toxins. Some research (Kooij, et al. 2016; Betts, et al. 2017) adds that the appendix might serve to promote beneficial bacteria in the gut. Australopiths had a large colon with gut bacteria to digest plant foods, with some means of food processing, like pounding (Newson & Richerson 2021). One point is that over the Miocene species diversity decreases and any early anthropoid diet, frugivorous with tender leaves, later becomes specialized, with digestive adaptations in the middle Miocene. In extant species we find that colobines are for leaves, hominoids for fruit, and cercopithecines for less ripe fruit.

Australopiths appear as the earliest hominins, circa 4.4–1.2mya. Among the oldest is *Ardipithecus ramidus*, a species of australopith (circa 4.5mya), and *Australopithecus anamensis* (circa 4.2mya), both of whom are mostly bipedal apes. Based on anatomy and dentition, *Au. anamensis* ate fruit and leaves, like chimpanzees, with some insects and possibly animal flesh (Welker 2017). *Ardipithecus* (some types dating to 5.8mya) possessed a hand close to a modern human's, but the foot differs from African apes and humans. Then, the long arms and short legs are not human, but apparent in older apes. This species appears to have been occasionally bipedal. *Ar. ramidus*, like *Australopithecus*, seems to have engaged in omnivorous eating, according to Francisco Ayala and Camilo Cela-Conde (2017). Humans with larger brains and small faces appear around 2.5mya. *Au. africanus* (circa 3mya) from the southern part of Africa had deciduous dentition, a combination of ape-human features, and preferred trees for feeding and shelter.

Other robust species (circa 1.9–1.2mya) preferred grassy vegetation. Let's consider an australopith v. an ape (Kerryn Warren, et al. 2019). For one, there is among australopiths bipedalism and different dentition, seen in reduced incisors and canines, compared to primitive chimps. Generally, an australopith had larger and flatter molars, thicker enamel, and a robust mandible to eat less hard plant tissue requiring incisors. The focus seems to have been on seeds, nuts, and like foods to crush and grind. Though, like chimps, australopiths preferred and ate fruit, their different dentition allowed for supplemental eating when fruit became scarce or was a competitive resource. There are nine species in four genera of australopiths: *Ardipithecus*, *Australopithecus*, *Kenyanthropus*, and *Paranthropus*. For all of these early hominins, the dental forms preclude reliance on meat or social insects, which need slicing power. Their dentition indicates crushing and grinding control for leaves, seeds, nuts, and such. As Klein (2009) says, for all these people, meat likely was not important before 2.6mya, shortly after which point rudimentary stone tools begin to appear. Of course sharp objects of wood, for digging roots and tubers, might have pre-existed stone implements, but none of those would have survived. Initially, stones could have been used opportunistically without any modification.

Stone implements to process meat seem to appear suddenly, but the tools were not immediately used for meat in these otherwise vegetarian species. While no one really knows, an educated guess posits that some tools of wood for root digging were already in use. With pressures for food, the sharp end of a stone by some individual was discovered to cut meat from a carcass. However, even if meat eating in the early *Homo* line is so archaic, that does not mean we can't evolve now and respond to the new selection pressures of bad health and climate change. In fact, we must. In competition for food, along with a drier climate and less vegetation to feed from, *Homo habilis* of 2.5mya began scavenging kills by big cats and in all probability used stones to de-flesh meat and smash bones for marrow. Similarly, today's environmental changes require that we adapt new methods of producing and using energy and food.

Fire does not appear until about 800kya, and it's not certain how much control over fire existed. One team of researchers (Hlubik, et al. 2019) date fire to 1.5mya based on fragments that "may be indicative" of cooked meats and plant foods. There's a valid question about how much raw meat early *Homo* could digest without getting sick, which might account for its low consumption. Moreover, truly effective group hunting among humans dates only to about 60kya. Primitive stone tools could also have been used to harvest vegetables, and living chimpanzees, among the great apes but not alone, are the most habitual and comprehensive tool users for food processing, like smashing nuts with stones or by using sticks to procure insects from holes.

Not untypically, biologists (Clark, et al. 2018) announce that the "true lines" of human descent are difficult to draw. For our purposes, outlines are sufficient. John Robinson (1985; Kerryn Warren, et al. 2019) confirms the large molars and thick enamel of *Paranthropus* for lengthy grinding of foods with reduced incisors and canines. For a biped, hands were used to process foods and for defense. For *Australopithecus* the molars are not as large, so there's less crushing with more importance to anterior teeth like canines and incisors. If the anterior teeth of *Australopithecus* were larger than those of *Paranthropus*, then they were probably used for offensive/defensive purposes or for feeding. Both species appear to be omnivores, feeding on vegetation and meat, says Robinson, but especially *Australopithecus* relying on meat during arid periods. Robinson sees *Australopithecus* in the line of tool users to come, since he says *Australopithecus* probably ate a "moderate amount" of meat. *Paranthropus*, on the other hand, was likely a vegetarian, thriving in a wetter climatic period and who evidences grit on dentition suggesting the consumption of root foods. As Klein (2009; Clark, et al. 2018) shows, there's no single phyletic australopithecine line but indications of how adaptive radiation follows new physical structures, such as bipedalism in the australopiths with dietary change to meat evident in later hominins tied to the rise of culture. What we term culture is a fast-moving artificial adaptation that slows natural selection and permits hominins to adapt in other ways, like the manufacture and use of tools, perhaps with nascent manifestations in *Australopithecus*.

In other words, *Australopithecus* in Robinson's (1985) thinking is more to "human" than *Paranthropus*, but because of close relatedness, "reversals" in evolution occurred and why there's some confusion about the evolutionary incongruity in the hominin line. This positioning is also reported by Kerryn Warren, et al. (2019). For Robinson, at bottom, *Paranthropus* is not a later development of *Australopithecus*. Clifford Jolly (1985) might see things differently. Concerning hominins, he focuses not on the *when* but the *why*. For example, rather than emphasizing bipedalism, small canines, tools, and larger brains, Jolly lays stress on larger molars and smaller canines as primate and not necessarily as hominin adaptation. The point of all of this, so far, is that no one can definitively say we evolved to eat animals or that we sprang up, like mythological Adam and Eve out of Eden, to rule the earth, hunt, and consume cooked, farmed meats. It just did not happen that way. Instead, there's a puzzling of different human/ape-like pieces that join, disassemble, and then reform in new manifestations. There's instability, a hallmark of metamorphosis in nature (Coccia 2021). One sees here adaptive variations in natural selection and descent with modification, recalling the evolutionary biology in Chapter 2. Included are other dimensions of evolution, particularly cultural, which will be detailed in due course in Chapter 5 as the logical sequence of this argument unfolds and concludes.

Thus, in line with our varied australopith ancestry that evolved feeding adaptations, we now need to resource more veggie diets in an effort to stem obesity rates and terminate global heating in a viable cultural ecology of plant-based foods. Tightening the focal lens more, let me look at some recent research in the next two sections.

Diet Variability and Adaptive Traits Like Dentition

Wataru Morita, et al. (2019) admit that molar variation in evolution between apes and humans is sketchy, so they compared enamel-dentine junctures between humans and extant hominoids like chimpanzees, bonobos, gorillas, orangutans, and gibbons. They learned that extant hominoids, but not humans, reveal a simpler alteration in molars and that humans have a more distinctive metameric shape, or linear series of segments, implying more modification over time. *Homo* has smaller molars as a "generalized form" (Ungar 2019). Dental reduction in *Homo* occurs after the Middle Pleistocene, a recent development arriving after the human-chimpanzee split and perhaps, as they say, because of neutral evolution (i.e., drift and not inevitably diet). This finding does not, however, dismiss adaptation through natural selection or negate the broad discussion of how early humans adjusted to an environment of evolutionary adaptation through food choices and sourcing. In fairness, I'm trying to cover all bases. In adaptation, tooth structure is selected for best use in eating sourced foods, says Peter Ungar (2019), an authority in this area. Foods have, he goes on, "mechanical properties" to which a species adapts, depending in part on how often these foods are consumed. Dental morphology is not always an indicator of preferred foods.

Francisco Ayala and Camilo Cela-Conde (2017) inquire about the adaptive traits of *Paranthropus* and *Homo*, derived from *Australopithecus*, enabling them to evolve in a cooling climate of increased savannas in the Upper Pliocene circa 2.5mya. The adaptation could have arisen through the *Paranthropus* robust mandible allowing for the consumption of tougher foods. Without that, *Homo* relied on primitive stone tools in an increasingly carnivorous diet mostly by scavenging. Primitive traits of *Australopithecus* can remain as a new species exhibits its own. For *Australopithecus*, initial molars; for *Paranthropus* there's a more robust masticatory muscle, producing a sagittal crest, a dimorphic trait mostly in males. By *Homo*, cranial capacity increases with a decrease in massive mastication. Some *Homo* species were probably not fully bipedal; that arrives in the Pleistocene, beginning circa 1.8mya. Without oversimplifying, early hominins like *Au. africanus* (circa 3mya) and *P. robustus* (circa 2mya) are two basic types, one gracile in form and the other robust in mandible.

The larger teeth of *Paranthropus* indicate vegetarianism. Robust australopithecines have thicker molar enamel than, for example, *Au. africanus* (circa 3mya), an omnivore whose diet included meat. This inference has been debated, according to Ayala and Cela-Conde (2017), where some say all the australopiths of South Africa relied on hard vegetation. Still others, apparently, based on microwear analysis, say *Au. africanus* was mostly a leaf and fruit eater. *Paranthropus robustus* (circa 2mya), in this examination, would have been more inclined to eat small and hard food items, like seeds. Whatever might have been, the *Au. africanus* diet was more varied than that of *P. robustus*, but as these paragraphs demonstrate, there is a mixture of adaptations and behaviors among early hominins. In fact, early hominins who foraged raw vegetable foods up to six hours a day achieved enough energy and calories for brain expansion, assert Alianda Cornélio, et al. (2016).

This confusion of evolutionary dialects helps the case presented here. No one can say with certainty that anatomically modern humans, in morphology or physiology, evolved directly from any exclusively meat-eating species, much less from a species in a linear progression. The picture painted here, from the paleontological evidence, colors our australopith ancestors, like our living great ape cousins, as more vegetarian than not, overall.

Some plant foods are chosen over others, which can determine a creature's habitat. Flexible choices are necessary because of scarcity and seasonal variation of food. *Au. africanus*, *P. robustus*, and *Au. sediba* (all circa 3–2mya) in the Cradle region of South Africa, say Amanda Henry, et al. (2018), among the earliest hominins chose different habitats and foods, from forest to savanna in a variegated landscape. This inference is based on isotope and microwear analyses. Cave sites with bones overemphasize food sourcing during dry periods. With reliance on herbivory, these authors say, varieties of landscapes might have affected hominin food preferences with some conscious tracking of plant protein, calories, and nutrition, seen in extant apes. Ian Towle, et al. (2019) actually studied dental caries, or tooth decay, in South African fossil hominins like *P. robustus*, *Homo neladi* (circa 335kya), *Au. africanus*, *Au. sediba*, and early *Homo*. Towle says *Au. africanus* ate large amounts of tough vegetation, lacks caries in fossils, and so was more of an omnivore. Yet, there's a significant dietary difference between this species and others, where caries are more common in *P. robustus* because of large molars and a diet that included fruits and honey. From the australopithecines to early *Homo* the fossil dental history exhibits a decrease in diet specialization, though some paranthropines are specialized and some not (Ungar 2019). Let's take an even closer look in the next section.

Microwear and Isotope Analysis

Although something has already been said about isotope analysis in passing, it might be useful to aggregate most of it into one section. While not repetitive, some of what's here looks at tooth wear literally from another angle. These researches are important for the overall argument in light of human prehistory and cultural evolution connected to food ecology.

Ayala and Cela-Conde (2017) report how in 2006 a team of researchers compared ancient *Homo* cheek teeth with extant primates, including contemporary hunter-gatherers. The conclusion is that the diets that specialized in hard items or tough foods have different wear patterns. Strontium/calcium rates have also been examined in tooth enamel. Leaves and grass have less strontium in contrast to fruit and seeds. This study, say Ayala and Cela-Conde, reveals that for *P. robustus* (circa 2mya) the strontium/calcium ratio is higher than for carnivores but lower than in the omnivorous baboon. *H. habilis* (circa 2.5mya) and *H. erectus* (circa 1.8mya) have strontium/calcium rates similar to baboons. Nevertheless, even for *H. habilis* fruit was an important dietary component, evidenced by his large teeth, though meat eating began to increase with tool use (Andrews & Johnson 2019). Additionally, herbivores who eat grass along with their predators will reveal higher carbon isotope rates. Ayala and Cela-Conde say that those who studied this compound conclude that *Paranthropus* fed mostly on low carbon isotope food, like fruits and leaves, or perhaps they sometimes preyed on herbivores who ate leaves. According to Barbara Welker (2017), isotope analysis of paranthropines in East and South Africa in the early to Middle Pleistocene reveals consumption of some "animal matter," although it could simply have been insects. Hominin reliance on meat does not imply obligate carnivory, Welker concludes.

For *Au. africanus* (circa 3mya) there's a higher carbon isotope ratio, suggesting a diet of fruits, leaves, and large amounts of grasses, or perhaps they occasionally preyed on animals with that diet. The dental morphology of this species from isotope and other analysis suggests sustenance variance because of seasonal and food variability, from fruits, leaves, and roots to grasses and sedges, but not revealed so much in other extinct hominins (Joannes-Boyau, et al. 2019). The implication is that some carnivory could precede *Homo* and tools. For *P. boisei* (circa 2.4mya) in East Africa, most of the diet, in the carbon isotope ratio analysis, reveals grasses and sedges of the savanna, like the diet of *Au. africanus*. Alexandria Peterson, et al. (2018) studied dozens of specimens of *Au. africanus* and *P. robustus* using microwear texture analyses that confirm some earlier studies. There was little difference between the two species, but *Paranthropus* had rougher and deeper dental wear features whereas *Australopithecus* had smoother and shallower surface wear. This suggests some habitat and dietary differences.

For some researchers, there seems to be an open question regarding any specialized *Paranthropus* diet (Kerryn Warren, et al. 2019). The South African diet of early hominins is complex and difficult to trace accurately, and there are conundrums. For instance, *P. boisei*, who specialized in C_4 foods blossoming in a hot, dry climate, disappeared when those forms of sustenance were prevalent. Research (Quinn & Lepre 2021) suggests a spike in C_3 vegetation, which favors a moist, cool climate with temperature transitions that decreased C_4 foods and instigated resource competition. In the end, Ayala and Cela-Conde (2017) say *P. boisei* was likely not a specialized eater but, like *Homo*, omnivorous out of necessity, since *Paranthropus*, by virtue of masticatory muscles and the isotope analyses, characteristically relied on large amounts of vegetation. However, reflecting the often inconclusive way to interpret such data, Bernard Wood (2005), a leading authority on human evolution, says the isotope analysis of the 1.5mya South African *Paranthropus* certainly indicates that he was partly a meat eater. In turn, this is questioned by Laura Martínez, et al. (2016) who say that the isotope data suggest a soft diet for *P. boisei* and less meat for *Homo ergaster* (circa 1.8mya).

Just to drive home the main point, these confusing and sometimes contradictory debates disclose how meat was not a predominant food source for our early ancestors or relatives who survived on a mixture of hard and soft plant foods, seeds, and nuts.

The South African *Au. sediba* (circa 2mya) reveals traits from other australopithecines and even *Homo*. From a few specimens examined by Amanda Henry, et al. (2012), there seems to have been a C_3 diet based on dental microwear and isotope analyses of leaves, bark fruit, vegetables, and water grasses or sedges. About 95 percent of earth's plants are C_3 and would include wheat, rye, oats, rice, cotton, sunflower, trees, shrubs, and some herbs, as opposed to C_4, which would include maize, sugarcane, switchgrass, and millet. For this species, says Henry, the C_3 foods were preferred, similar to *Ar. ramidus* (circa 4.5mya) and extant chimpanzees of the savanna though other plants like sedges were available. Evidence from isotope analyses, Henry goes on to say, reveals that *Australopithecus, Paranthropus*, and early *Homo* diets differed from modern chimps and the *Au. sediba* specimens examined. There are molar microwear differences between the specimens with one suggesting the consumption of hard foods more than seen in, for instance, *Homo erectus*. Isotope analysis shows these *Au. sediba* people ate what savanna chimpanzees eat today, and an early species close to the diet of these specimens would be *Ar. ramidus*.

Isotope studies reveal that C_4 foods were consumed by hominins relative to the C_3 foods of other African apes, depending on environmental factors and availability, attesting to the flexibility and complexity of diets (Wynn, et al. 2020). Jonathan Wynn's work consisted of a carbon isotope study revealing C_4 foods of *Paranthropus* in Ethiopia with a shift to C_4 foods circa 2.3mya followed by other hominins in this "fossil sequence." However, in this same Ethiopian area other hominin diets differed because of varied food sources. In other words, diet in early people was a response to and a relationship with the environs. I'm making an argument that modern humans, too, must now respond to the environment with a new diet. Because of microwear and isotope analyses this means not all paranthropines, says Wynn, thrived on hard objects (*P. boisei*, circa 2.4mya) while others did (*P. robustus*, circa 2mya).

Thus, there's no common adaptation because of geographical differences. *P. boisei* is one of those non-ancestral relatives who lived among our early *Homo* ancestors in east Africa. This species, say Bernard Wood and David Patterson (2020) had very large post-canine teeth with extremely thick enamel. They ask why a hominin would need such a strong mandible and large teeth. They admit there's a "dietary puzzle." While the theory is that this species with this morphology in a cooler climate was eating more grasses and sedges, they see such chewing strength as evidence of nut and seed processing, food supplements. Wood and Patterson conclude that there was a "complex relationship" in response to environmental change and diet for some species.

Jordi Marcé-Nogué, et al. (2020) also admit there's debate and contradiction about early hominin feeding because inferences appear from different methods. They used a sample of thirty extant primate species for a comparison with hominin fossils to analyze chewing biomechanics. They conclude that *Paranthropus*, otherwise known as "nutcracker man," relied on more soft rather than hard foods, consistent with recent microwear and isotope studies, so no adaptation for hard foods. Indeed, Ian Towle, et al. (2021) studied *Paranthropus* fossil literature in relation to extant primates, analyzing tooth fracture severity and position. They conclude that *P. robustus* and *P. boisei* (circa 2.4–2mya) did not habitually consume rigid foods, but ate more C_4 grasses and sedges with a low chipping rate similar to gorillas, chimpanzees, and gibbons, but not orangutans who eat hard seeds and nuts. As with extant primates, a species that evolved to eat soft foods could still eat stiff items depending on season. In *Paranthropus*, the evolution of thicker enamel might have been an adaptation for greater wear and not necessarily to consume hard foods, considering the rarity of chipping, says Towle.

Adam van Casteren, et al. (2020) also admit that the discussion about early hominin diets is "heated." Nanowear experiments, they say, show very hard shells cause little damage to enamel, so no enamel pitting. Small, but hard, seeds from grasses and sedges might have been crucial early hominin foods. Most hominin foods from 3.5mya forward, they say, based on isotope evidence, were mostly C_3 vegetation with a good amount of C_4 foods. Van Casteren advocates for a diet of seeds, compared with others who argue for leaves or tubers. Australopith dental-cranial mechanics could generate a high bite force. The strong enamel and low, blunt cusps of australopith molars would present fracture. Generally, microwear patterns on teeth of Plio-Pleistocene hominins beginning at 5mya onward reveal a low to moderate striation or scarring, except for *P. robustus* (circa 2mya), says van Casteren. Dense, woody plant tissues create enamel marks but not fissures, which could come from grit in grasses. These authors conclude that energy-rich C_4 grass or sedge seeds were an important nutrient for many australopiths.

Other researchers have studied tooth fracture. Gary Schwartz, et al. (2020) say relative enamel thickness is a standard metric. Thick enamel makes crowns strong enough to resist fracture from hard foods. Schwartz offers a new metric of "absolute crown strength" to evaluate dental structures resistant to fractures in fossils and living hominoids. Generally speaking, the robust australopiths (*P. robustus* and *P. boisei*, circa 2.4–2mya) had stronger molars than their gracile counterparts (*Au. anamensis*, *Au. afarensis*, and *Au. africanus*, circa 4.2–3mya), where molars are yet stronger than those in chimps. Gorillas have, broadly speaking, less thick enamel than orangutans, though gorillas crack nuts with their teeth. Simply put, enamel thickness and absolute crown strength need to be considered together. Living apes, modern humans, and australopiths have different areas of strength and weakness in crown fracture. From the analysis by Schwartz, it seems strength against margin fractures at the crown cervix, as opposed to a radial median fracture at the cusp tip, might be an adaptation derived in chimpanzees, gorillas, and the last common ancestor for eating soft, compliant, and tough foods. Hard items might not have been as crucial to feeding. Strong cusp enamel would have benefited eating hard objects indirectly, but this wears out first and why reliance on relative enamel thickness could be inaccurate by making one believe hard foods were more important to australopiths than to orangutans. Schwartz says cusp enamel forms first and is thick, followed by lateral enamel, but admits these findings are inconclusive.

Once again, this jumble demonstrates how early hominin diets were not restricted to meat; animal flesh was not an essential component of their diet. Conflicting ideas do not weaken the argument here, for two reasons. 1. Despite these debates about dentition, microwear, etc., the consensus points to a plant-based diet. 2. Our ancestral relatives and distant kin evolved biological adaptations as contemporary humans must evolve cultural adaptations about food ecology. Knowing history can be an important guide on the path forward.

Let's try to bring this section to some conclusion.

In terms of the hominin evolutionary lineage, some paleontologists lump species together in one line while others split them into parallel lineages. This adds to the confusion. One observation seems sure: there are at least two early hominin australopith lineages, gracile and robust (Kerryn Warren, et al. 2019). That is, *Au. africanus* (circa 3mya) because of its special mastication gives rise to ape-like cranial and dental traits of *P. robustus* (circa 2mya). In this clad (family relationships) is the even more specialized *P. boisei* (circa 2.4mya), one of many likely vegetarians, with even larger masticatory development. Some think that the robust nature of mastication follows a logical pattern. This trend (i.e., *Au. africanus* inclining to *P. robustus* leaning to *P. boisei*, circa 3–2mya) might appear too simply limned for some paleontologists. One could argue that hominin evolution is unkempt in its history through the paraphyletic *Australopithecus* making it difficult to pinpoint phylogeny (time relationships).

In any event, Ayala and Cela-Conde (2017) place *Homo* as the sister group of *Paranthropus*. Then there are the changes in the cranial and dental features of *H. habilis* (circa 2.5mya) from *Australopithecus*: smaller mastication structures like molars, thinner dental enamel, and a larger cranium. For *Homo*, the distinguishing feature from others and any predecessors is that which is external, i.e., tool manufacture to scavenge and possibly use to hunt in a very primitive manner. Put differently, a valid question might hover around whether or not *Homo* could have continued the mostly vegetarian line of the predecessors absent tools. In a temperate climate, *Au. africanus* (circa 3mya) survived into adulthood better than *H. habilis* in a cooler environment. This raises a question about a selection pressure for tools and meat. The point, precisely, is that modern, industrialized humans are experiencing pressures that pose risks to which we must adapt by shucking away older systems of eating for the food ecology of veganism.

Ayala and Cela-Conde (2017) suggest that *P. boisei* (circa 2.4mya) could plausibly have made tools, as we see in extant chimpanzees. However, at this time it's not known that chimpanzees use stone tools in hunting, yet they commonly throw rocks. Although infrequent, sharpened sticks, on the other hand, can be used by chimps to spear bush babies from trees. Compared with *Australopithecus*, where some researchers would place *H. habilis* (circa 2.5mya), *habilis* is more gracile in cranium, mandible, and post-crania. Incisors are larger than, though molars overlap with, *Australopithecus*. There is also similarity in some features from *H. erectus* (circa 1.8mya) to *H. sapiens* (circa 250kya), reflecting the thrust of the evolutionary medley. All this is magnified by variation among *H. habilis*, too, as seen in the adapted morphology, physiology, and behavior of current ape species and even among ape groups. For sure, every individual modern human varies dramatically in strength and size, regardless of sex.

An evolutionary relationship can be posited among *Ar. ramidus* (circa 4.5mya), *Au. anamensis* (circa 4.2mya), and *Au. africanus* (circa 3mya) but, as Ayala and Cela-Conde (2017) point out, there's no consensus on the hypothetical transition from *Ar. ramidus*, *Au. anamensis*, and *Au. africanus*. Multiply that discord among the many tracks possible from early hominin fossils where, over time, populations share alleles, gene forms on a chromosome subject to mutation. *Au. anamensis* is the first known hominin to eat tough, vegetable foods, followed by *Au. afarensis* (3.9mya), who has an increase in mandible robustness, possibly feeding from soft foods but potentially folivorous feeding on tough plants and grasses. There's no definitive answer. Frankly, this muddle helps prove, in clarified thinking, that humans can and should adapt their diets according to environmental stress. Species like *Au. anamensis* survived by eating nuts, seeds, and other hard foods, and there are implications from microwear analysis that their consumption implied fibrous foods similar to chimpanzee and gorilla diets. Microwear analysis reveals, likewise, in *Au. afarensis*, a diet of tough fibers. Eating tubers and roots from the earth implies grinding sand on the top surface of the tooth, Bernard Wood (2005) relates. People like *Au. afarensis* were habitually bipedal and had primitive traits for arboreal behavior, perhaps for protection and leaf eating, since hands could be used to manipulate leaves for food as well as for nesting. The jaw morphology of both of these species (*Au. anamensis* and *afarensis*) indicates a fruit diet with fallback to hard resources when fruit became seasonally scarce. Today, we see that dietary behavior in our orangutan cousins, fruit eaters, who rely on tree bark during stressful times. Were they still extant, we'd see how hominins descended from these "apes," though not in any simple linear fashion as popular media would suggest.

For example, *Au. afarensis* (circa 3.9mya) might be ancestral only to *Paranthropus* and not to *Homo*, the kind of mismatch matrix, says Klein (2009), alluded to in this chapter. Yet the organization of the *Au. afarensis* brain from endocasts is apparently human, seen in *Au. garhi* (circa 2.5mya). Unlike other australopiths, *Au. africanus* (circa 3mya), although chimp-like, shares cranial and dental traits with living *Homo*. Post-cranially this is an ape-like hominin except for a human-like foot. This species had short canines with little sexual dimorphism. *Au. africanus* and specially *Paranthropus*, compared with *Au. afarensis*, ate more hard vegetation like fibrous tubers and nuts. There was great diet variability and overlap with these two species. Their enriched dental enamel suggests a diet of grasses and insects thriving in grasslands. Later, *H. ergaster* (circa 1.8mya) would feed on a variety of foods, yet with stone technology probably consumed more meat and marrow, since he possessed no meat slicing cheek teeth.

P. boisei and *robustus*, both australopiths (circa 2.4–1.2mya), east and southern Africa, overlapped with early *Homo* around 1.8–1.5mya. *Paranthropus* was more of an herbivore than a carnivore, which allowed her to flourish. *Paranthropus* is a special offshoot of hominins. *P. boisei*, in comparison to other australopiths, though somewhat ape-like, had a more enlarged brain paralleling *Homo*. She had large cheek teeth molars but a small body with reduced canines and frontal teeth, so less ape-like in those features than even modern humans. *P. robustus* was likely an exclusive vegetarian and ate subterranean foods like roots and tubers, along with a combination of grasses and seasonal non-grassy foods like seeds. Insects on this vegetation were probably consumed, or separately, like termites, as seen in extant chimpanzees and hunter-gatherers. The less specialized dentition of *P. robustus*, compared with *Au. africanus* (circa 3mya), implies omnivory. Microwear analysis of *P. boisei* fossils shows no sign that they dentally crushed bones.

In sum, our early hominin ancestors and other relatives were not obligate meat eaters and, likely, more vegetarian (as earlier defined in Chapter 1) than anything else. We evolved from this array of plant-eating species and yet have somehow convinced ourselves that, like big cats, we are dedicated carnivores and need to consume massive quantities of meat. Nothing could be further from the truth, as you have just read.

Australopiths as Prey, Not Hunters

While some of this chapter's content might seem confusing, it's important in making a case for the evolution of veganism to illustrate exactly how modern humans evolved in a kaleidoscopic manner from several related and ancestral species who did not rely on meat as a staple food. In fact, some scholars suggest that contrary to the media-popular and masculine notion of the hunter, early hominins were prey of many other animal species (Newson & Richerson 2021). This revelation puts an added twist to our lineage as model omnivores, the evolution of our vegetarian ancestors, and the claims in a case for veganism.

Anthropologists Donna Hart and Robert Sussman (2005), for instance, argue that australopiths were ready food for a host of predators, from big cats and hyenas to crocodiles and raptors. Orangutans, gorillas, and chimpanzees among many primates are preyed upon (Strier 2017; Jaffe 2019). Even contemporary human babies and adults are victims of nonhuman predators. There's no surprise in, and a long history of, humans as food for big cats, bears, wolves, giant snakes, large birds, and huge reptiles. We act surprised, but that's because we consider ourselves the apex meat eaters. If humans are prey food now, they certainly were in prehistory with their smaller bodies, and so the notion of "man the hunter" is not fully accurate. Some years ago, a fast-food burger chain ran an advertisement called, "Where's the Beef?" According to Hart and Sussman, early hominins were the beef in prehistoric times.

Humans evolved in an environment where they were under attack as meat, opposed to regularly hunting for and eating animal flesh, say Hart and Sussman (2005). Holes in, and scrapings on, australopith skulls of children and adults match leopard fangs and eagle claws. Skull holes in early *Homo* at Dmanisi, Georgia 1.7mya indicate predation by a saber-toothed felid. Apparently, deadly encounters between early humans and big predators would have occurred at night, as is the case of nighttime field observations of baboons who become prey. Hart and Sussman see brain growth driven by humans as quarry, not in their eating of meat. Australopithecine fossils are not like those of carnivores. They don't have the claws and fangs of predators or even tools of the hunt. Early twentieth century marginal theories up to the 1950s painted australopiths as, literally, wolf-apes hungry for meat, say Hart and Sussman. This perverted attitude climaxed in Robert Ardrey's popular 1961 book *African Genesis* that purports how early hominins were killer hunter-apes. The reality is that leopards were attacking these small, early humans by biting into their skulls, as research on the puncture holes shows when compared to fangs.

Unfortunately, this dogma of a small bipedal ape as a machine evolved to kill and eat meat was thrust to the forefront with one of the world's most prominent paleoanthropologists, Louis Leakey, say Hart and Sussman (2005). They go on to recount how gorillas and chimpanzees are preyed upon in Africa by leopards, lions, or hyenas and orangutans in Asia by leopards and tigers. They note in particular how in Parc National du Niokola Koba, West Africa, chimps there usually do not engage in hunts for monkeys since they themselves are preyed upon by carnivorous big cats and wild dogs. Curiously, this data contradicts what others generalize in Chapter 3 about chimps as meat hunters, you might recall. In the Mahale Mountains, during one research study, say Hart and Sussman, 6 percent of chimpanzees fell prey to lions each year, with similar chimp predation rates, regardless of age or sex, in the Tai Forest, Côte d'Ivoire, West Africa. However, as they indicate, there's a population balance between prey and predator that only gets offset with human incursion into resources or reduction of habitat size, as already noted, but worth repeating. You can see how any depiction of our cousin chimps as crazed meat eaters is inaccurate, as demonstrated in Chapter 3.

In leopard diets alone, from Africa to Asia, primates can make up to 80 percent of consumed food. This coevolution of predator-prey selection pressure is crucial to Hart and Sussman (2005; Strier 2017). They argue australopith and other early human adaptations to survive predation, including group cooperation, increased brain size in such a gracile species, not meat eating. Of course, a combination of factors affecting cognitive, intellectual, dietary, and social development is more likely. Nonetheless, their research shows that all primates are preyed upon by any number of carnivores, and since humans are primates and were small between 5–2mya it stands to reason that early people of all ages day or night endured pressure in an environment of evolutionary adaptation as subjects of predators. This deduction has to be put next to the inflated emphasis on small groups of chimps in some confined areas who prey on a few monkeys any given month. In the meantime, Hart and Sussman estimate that the predation rate on primates is as high as 25 percent. Karen Strier (2017) similarly has much to say about prey/predation among primates and other species. There's a constant risk that affects primate social, foraging, and reproductive behaviors. Mouse lemurs suffer a high rate of predation but reproduce much more than great apes. It's already been suggested that ape birth intervals might be an adaptation not just to parenting but also to population control relative to maintaining ecosystem harmony. That idea does not mean great apes did not reproduce. Rather, there were a number of selection pressures operating simultaneously, of course. Like humans, ape mothers invest years of time and energy in a single child. According to Hart and Sussman, as evidenced in primate fossils, up to 10 percent died of predation, including more evidence from the Pliocene and Pleistocene, with many large cats with saber-like teeth that exact a downward knifing motion.

We evolved to avoid many predators, not strictly to eat meat. Most primates are tree dwellers, and that's how they avoid predation. Early hominins like australopiths were still forest beings but found other ways to deal with predation. There are many modern stories up to this day of people as prey to wolves, bears, or big cats. However, in an open savanna environment with some tree protection on the edges, there were many prey and predators, ample carcasses for scavenging. Bipedalism to investigate meat sources also freed the hands to work, signal with vocalization, carry, and allow for a cooler body temperature. There was, over evolutionary time, human growth in morphology, physiology, and brain size as savanna scavengers and gatherers. In short, Hart and Sussman (2005) say we evolved not as hunters, but to avoid violence inflicted on us.

Bonobos, gibbons, and baboons are known to hunt small mammals occasionally, not just chimpanzees, but rarely orangutans. Male chimpanzees in Gombe can kill up to 30 percent of colobus monkeys per year. Colobus do not have a high reproduction rate to offset this predation. Importantly, Hart and Sussman (2005) say this predation is recent, perhaps due to habitat loss and food resources depletion across species because of human invasion. They note, incidentally, by analogy how Jane Goodall (1986, 1971) unwittingly triggered aggressive group behavior among chimpanzees by setting up a feeding station. Goodall later corrected this. Colobus monkeys could not have survived if they were exclusive meat for chimps. This is asymmetrical predation, not population balance.

Perhaps human incursion has limited chimpanzee domains too much. I don't mean to harp on this point, having raised it in the section on chimps, but it's important in the total argument about selection pressures and the cultural ecology of food. Meat, though prized by chimpanzees, is not the preferred food. That's apparent in human history, too, and for the sake of unhealthy pandemics and climate change we might want to meditate on this fact of life.

Early Homo Species

Each ape species has evolved to fill an ecological niche, hearkening back to the biology of Chapter 2. Fossil evidence of early hominins like *Australopithecus* and *Paranthropus* rules out meat-based diets. Australopiths survived for a long time, exploiting lots of resources on the savanna, on the edge of the savanna/forest, and in the forest. In our own time without vegan cultural evolution, resources will become depleted, whether water, energy producing minerals, or the animals who maintain ecosystems on land or in water. Since primates, including humans, are not obligatory carnivores by nature, we should not assume that raw meat was regularly consumed. Early tools were used for other foraging purposes. Even among extant apes, rarely are tools used to hunt, except perhaps in the isolated spearing of bush babies. More routinely, tools are used by living great apes to process plant foods, nuts, or insects.

Looking past the early hominins, let's say a few words about our more direct ancestors who factor into the larger picture here but not quite in this argument. *Homo habilis* (circa 2.5mya) had a larger cranial area and smaller cheek teeth than *Australopithecus* and *Paranthropus*, though there was variation in the evolutionary matrix with some specimens having teeth like *Australopithecus*. Later, *Homo erectus* (circa 1.8mya) had smaller teeth and a larger cranial capacity. *H. habilis* and *rudolfensis* lived among *P. boisei* and *robustus* around 2–1.2mya. *H. ergaster* appears in eastern Africa 1.8mya, perhaps deriving dentition from *H. habilis* and a cranium from *H. rudolfensis*, but this is unclear, says Klein (2009; see, too, Clark, et al. 2018). Brain encephalization could have emerged independently. That is, shared genes of similar species increase the likelihood of comparable adaptations to similar selection pressures in analogous environmental circumstances.

The early hominin australopiths were followed by *H. erectus* who thrived for a very long time as a chronospecies, namely, from *H. ergaster* and other relations in Africa to a larger brained *H. erectus* in Eurasia (Kerryn Warren, et al. 2019). Then, by about 500kya from the ergaster/erectus sphere (with some species on the margins) there is *H. heidelbergensis* with a much larger brain and more culture. *H. erectus* lived until as late as 60kya and in one form 12kya as *H. floresiensis* on Indonesian islands, so a very successful species. We have not yet lived as long. From these archaic humans Neanderthals arise in Europe during the Ice Age. Far in the deep south of Africa another archaic human evolved as *Homo sapiens*, our species, with a larger brain and a more gracile body. Robin Dunbar (2014) notes that while there were many female *H. sapiens* breeding, all living human beings can trace their mitochondrial DNA to about 5,000 women from around 200kya.

Neanderthals and Homo Sapiens

In Lower Paleolithic people like Neanderthals and their predecessors *H. heidelbergensis* and *H. erectus*, circa 2mya to 250kya, there would have been seasonal ecological and social strategies for food gathering and processing. Theirs was a generalized diet, says Robert Hosfield (2020), open to continuing debate about levels of plant v. animal foods. He notes, for instance, that Neanderthals ate seeds and leafy plants, and not only meat, eggs, or marine foods, to supply essential fatty acids. Energy was needed to capture spotty prey. Most Lower Paleolithic areas occupied were temperate woods offering abundant plant foods, so meat was not necessarily a daily experience, says Hosfield. Researchers (Fellows, et al. 2021) found amylase enzyme residue on Neanderthal teeth indicating a boost in sugars from starchy foods and hence a diet dependent on plants. In Paleolithic times, plant foods amounted to a "significant percentage" of the human diet (Nowell 2021, p.143). The typical viewpoint, however, is that the Neanderthal diet was heavily geared toward red meat though raw and cooked vegetables were readily consumed based on dental analysis (Paskey & Cisneros 2019).

Regardless, there would have been food sharing networks. One interesting note from Hosfield (2020) is how Neanderthals might have eaten rotten meat stored underground or in water as an alternative to cooking. This practice of preservation and preparatory digestion, although offensive to most modern humans, demonstrates how there is a cultural ecology of food that can change. Similarly, with scarcity of seasonal plant food in colder climes, these people, Hosfield notes, engaged in gastrophagy or the eating of partially digested remains in an animal's stomach. In some hunter-gatherer cultures this method of eating predigested foods is still practiced. This procedure would be unacceptable to some societies today, bolstering my point about the cultural ecology of food. Energy and time are required to find and prepare meats, as in cooking; instead, pounding of plant foods would have been more easily employed. Kiss feeding predigested food by the mother to her baby is less stressful in terms of milk production and offers good bacteria to the child (Newson & Richerson 2021).

Nutrition flexibility even among plant foods would have been essential to Neanderthals, though in many locales dependency was on the meat of ungulates. Diet is related to many behaviors, including sociality and migration. Like modern humans, weaning for Neanderthal infants began at five to six months, say Alessia Nava, et al. (2020) based on isotope technology of deciduous teeth. Good nutrition was required for the establishment of tooth enamel. Some herbivore environments of Neanderthals no longer exist, and they were surely good hunters and not simply scavengers. Neanderthals, generally, were not specialized to a particular environs or taxa, says Robert Power (2019), but lived on a subsistence strategy of the best foods, counting many plants but still animals. Isotope and other analyses, Power goes on, reveal only high protein food, like meat, but present at best a "generalized picture" of cultural food ecology. Correspondingly, Hosfield (2020) says that recent research using isotopic and dental analyses reveals less reliance on protein rich animal diets among Neanderthals, especially in comparison to current hunter-gatherers. These conclusions vary, contingent on the time of year, geographical location, and mobility. Yet evidence reveals, Power stresses, that Neanderthals ate nuts, seeds, olives, roots, berries, lentils, peas, vetchling, grass husks, legumes, fruits, etc. Depending on plant food concentrations in warmer areas there was processing, but apparently there is confirmation of abundant plant use even during cold periods.

Because of living in higher latitudes and with a cold climate, Neanderthals collected meat from animals as a major food source. There's no doubt, however, say Christopher Stringer and Clive Gamble (1999), that there were copious amounts of plants readily consumed, too. For example, these authors note, in Germany around 200kya there were temperate zones and warm springs around Weimer that encouraged a flourishing of abundant flora. Neanderthals had no real home camps until about 60kya with, for instance, stone hearths, post holes for coverings, pits, trenches, etc. This lack of a base site suggests regular movement or no real importance assigned to any particular butchering, cooking, or stone knapping plot. Stringer and Gamble are skeptical of regular, specialized big game hunting by Neanderthals that is still promulgated (Paskey & Cisneros 2019). It seems that small animals were preferred. They also note that scavenging as well as hunting is not irregular behavior but involves planning in terms of knowing where to go at which time and involved stone technology.

Rather than focusing on bones at Neanderthal sites, which could have been hunted, scavenged, or left by big cats or hyenas, it's more productive to consider all the elements of various Neanderthal regions with "subsistence security" (Stringer & Gamble 1999) based on resource strategies dependent on the flora and fauna of any province. For instance, Laura Weyrich, et al. (2017) show that DNA evidence culled from a cave in Belgium, a cavern in Spain, and a grotto in Italy, disclose there's no surprise that, generally, Neanderthals varied their diets based on regional ecology. These samples reveal lots of meat eating, and although not pumped up with antibiotics or hormones, this was not an ideal diet (Cordain, et al. 2000). However, from a cave in Spain, there is no evidence of meat eating and, instead, mostly mushrooms, pine nuts, moss, and forest gatherings. Microwear analyses from other ecological areas, says Weyrich, indicate diets centered on what was available, like plants. As noted above in terms of great apes and australopiths, sustainable diets were based on variety, not on meat alone. Oddly, today, many modern humans lavishly consume processed meat, eating it three times a day, filling their bodies with excess fats, too much protein, sodium and nitrates, and unnecessary growth hormones and antibiotics. Since meat and dairy were not exclusive foods for our hominin ancestors, we don't need to consider those corporate agricultural products as sustainable food ecology today.

H. sapiens spread quickly and widely replacing other hominins and eventually Neanderthals. Some of these displaced hominins, to be clear, had been around for many hundreds of thousands of years. By 70kya modern humans were out of Africa to Asia and, remarkably, crossed a sea to Australia about 40kya. By 50kya much more sophisticated tools, weapons, jewelry, and art began to appear. For a full discussion of Paleolithic material and art culture, see Tague (2016, mainly 2018). At the Levant in the Middle East, modern humans probably had contact with Neanderthals, but the modern humans detoured to Asia, where they met remnants of *H. erectus* and Denisovans, about 41kya, with whom they bred.

Denisovans are genetically connected to Neanderthals through some ancestor and are the last of archaic humans who migrated eastward. *Homo longi*, or Dragon Man with a considerable skull and equally massive molars, is the only Denisovan cranium found, though some researchers (Ji, et al. 2021; Ni, et al. 2021) claim him as a new hominin species, so named, because of his combined mosaic of human features. The diet and tools of *H. longi* are unknown, although he lived during an interglacial period with frigid winters, if residing in that Harbin region of China during cold months. By 40kya modern humans were in Europe and slowly displaced Neanderthals, who evolved there from about 250kya. Modern humans and Neanderthals lived together until about 28kya when Neanderthals vanished. Just as archaically modern humans spread widely, rapidly, eating and taking all in their path, so too did they replace Neanderthals.

This observation is based on the aggressive and acquisitive behavior of humans from early history (i.e., the agricultural age of circa 10kya) to the Middle Ages and thereafter. Ignoring the viciousness of sanctions, rival trade, aggressive wars, computer hacking, ransomware, etc., Steven Pinker (2011) claims better angels of human nature have harnessed a decline in violence and so might disagree. Pinker-like thinking ignores the billions of animals caged, farmed, and slaughtered for food yearly, as if that's not violent. At present, humans are working to displace primates in tropical forests. This is not to say, coming back to the related point, that there was no communication with Neanderthals. Anatomically modern humans and Neanderthals interbred, and many people of European descent have remnant Neanderthal genes. It cannot be denied that Neanderthal extinction was followed by *H. sapiens* replacement (Hublin 2017). Neanderthal populations were small, challenged by the environment, and included inbreeding (Vaesen, et al. 2019), so their demise could have come from a number of factors. Competition avoidance prevails among hunter-gatherers, so a theory of non-violent destruction is tenable but ignores other forms of survival behavior. Some might take great lengths to secure a food source that's shared with the same or another species.

Circling back, all told, the data from this chapter shows that anatomically modern humans are not born hunters and eaters of meat. Rather, "hunting" is indicative of behavior we see in most industrial societies today: forceful depletion of resources. As revealed, our australopith prehistory is not one of excessive consumption, and there was little meat eating, if any, in some species. Similarly, later in the hominin evolutionary line, Zink and Lieberman (2016) insist that with smaller teeth, bite force and chewing time decreased advantageously in *H. erectus* by using stones to smash tubers and mechanically process small portions of meat before the advent of cooking. Bonnie Yoshida-Levine (2019) bluntly states that *H. erectus* was not just a meat eater. The effort involved in fashioning and employing such tools for food production would have necessitated a great advantage if other energy and protein plants were readily at hand.

No matter, since food, including how it's prepared and eaten is a cultural enterprise among humans, precisely the emphasis in this argument about the evolutionary case for veganism. No one would dispute the benefits of cooked meat in early human diets, but as this brief survey suggests, it could have been as much nutritional as well as a cultural commodity and bonding mechanism. Today, with the recognition of goals for maximum health and climate sustainability, a vegan culture can become the norm. How often do we hear the expression, *food is culture*? Let's prove it and enhance food ecology to ensure our healthy survival and environmental stability. In the process, a bonus is that we will have spared every year pain and suffering for billions of farm animals destined for human stomachs.

As you can see from the previous chapter and this one, our mostly vegetarian australopith ancestors and extant great apes thrived among each other and other species peacefully, but of course there are some exceptions. Because Neanderthals and Denisovans lived in mostly chilly and not tropical climates, unsurprisingly they typically consumed more meat, but were wisely in tune with the nutritional benefits from vegetation, seeds, nuts, etc.

Diet Hybridity and Cultural Evolution

Since this chapter's remit deals only with early hominins, the story ends here, with the first known use of stone tools for meat processing among *H. habilis* and *ergaster*. G. Philip Rightmire (1990) says plesiomorphic characters (shared evolutionary traits through similar ancestry) make diagraming *H. erectus* ancestry difficult in drawing even a jagged line from *H. habilis* or australopiths to *H. erectus*. The contorted lineage, in spite or because of adaptation, shows family relationships to earlier frugivorous and folivorous hominins. Rather than a graded sequence with chronospecies (derived), Rightmire sees *H. erectus* as a true paleospecies (ancestor status). That conclusion, however, does not exclude any evolutionary relatedness to other species, as indicated by Klein (2009), Ayala and Cela-Conde (2017), and even Mary Ann Clark, et al. (2018) or Bonnie Yoshida-Levine (2019). Understandably, there's no direct transformation of one species into another, *H. erectus* into *H. sapiens*. Rather, there are many hybrid states and offshoots between and on the edges in agreement with Darwin's (1859, 1871) descent with biological, ecological, and cultural modification. That observation helps bolster the argument of this book, since we are not programmed to be city dwellers, office or factory workers, any more than we are existentially predetermined to eat meat and dairy. In fact, by virtue of cultural evolution, we can and should alter practices like the farming and eating of animals that put our future at risk.

There were morphologically and physiologically different populations from the *H. sapiens* clad across Africa, say Eleanor Scerri, et al. (2018). This implies that various hybrids (e.g., *H. heidelbergensis*) resourced a medley of foods, like earlier australopiths. This human diversity was formed by ecology, as with earlier species. The roots of *H. sapiens*, says Scerri, are deep, with independent evolution in various African locales. There were multiple groups, separate but linked, each evolving different characteristics. This regionalization does not mean all evolving early modern humans were strictly meat eaters, though of course the tools of some indicate opportunistic carnivory. Habitat variability and environmental factors suggest, as seen in extant primates, a mixed diet. Tools across groups, as with physical traits, evolved independently and then advanced when groups met and exchanged cultural artifacts and ideas.

All of this prehistoric thinking leads us into the next chapter on contemporary people, us. We'll delve more deeply into cultural evolution, building from the biology, diets, and evolution of primates, great apes, and early hominins covered so far. As you read, keep in mind what has been implied all along: as biological organisms, creatures do not subsist in a vacuum, but eat in an ecology that affects other living beings in a habitat.

References

Andrews, Peter. 1985. "Species Diversity and Diet in Monkeys and Apes During the Miocene." *Primate Evolution and Human Origins*. Russell L. Ciochon and John G. Fleagle, eds. Menlo Park, CA: Benjamin/Cummings Publishing. 194–204.

Andrews, Peter and R.J. Johnson. 2019. "Evolutionary Basis for The Human Diet: Consequences for Human Health." *Journal of Internal Medicine* 287 (3): 226–237. doi:10.1111/joim.13011.

Ayala, Francisco J. and Camilo J. Cela-Conde. 2017. *Processes in Human Evolution: The Journey from Early Hominins to Neanderthals and Modern Humans*. Second edition. Oxford: Oxford UP.

Betts, J. Gordon, et al. 2017. *Anatomy and Physiology*. Houston, TX: OpenStax/Rice University.

Boesch, Christopher, et al. 2020. "Chimpanzee Ethnography Reveals Unexpected Cultural Diversity." *Nature Human Behavior*. https://doi.org/10.1038/s41562-020-0890-1.

Clark, Mary Ann, et al. 2018. *Biology 2e*. Houston, TX: OpenStax/Rice University.

Coccia, Emanuele. 2021. *Metamorphoses*. Robin Mackay, trans. Cambridge, UK: Polity Press.

Cordain, Loren, et al. 2000. "Plant–animal Subsistence Rates and Macronutrient Energy Estimations in Worldwide Hunter-gatherer Diets." *American Journal of Clinical Nutrition* 71: 682–692.

Cornélio, Alianda, et al. 2016. "Human Brain Expansion During Evolution is Independent of Fire Control and Cooking." *Frontiers in Neuroscience* 10: 167. doi:10.3389/fnins.2016.00167.

Darwin, Charles. 1859. *On the Origin of Species*. Joseph Carroll, ed. Ontario, CN: Broadview P. 2003.

Darwin, Charles. 1871. *The Descent of Man*. London: Penguin Books, 2004.

Dunbar, Robin. 2014. *Human Evolution*. London: Pelican.

Dunmore, Christopher J., et al. 2020. "The Position of *Australopithecus sediba* Within Fossil Hominin Hand Use Diversity." *Nature Ecology and Evolution* 4: 911–918. https://doi.org/10.1038/s41559-020-1207-5.

Fellows, James A., et al. 2021. "The Evolution and Changing Ecology of the African Hominid Oral Microbiome." *PNAS* 118 (20): e2021655118. https://doi.org/10.1073/pnas.2021655118.

Goodall, Jane. 1986. *The Chimpanzees of Gombe: Patterns of Behavior*. Cambridge, MA: Harvard UP.

Goodall, Jane. 1971. *In the Shadow of Man*. Boston: Mariner.

Hart, Donna and Robert W.Sussman. 2005. *Man the Hunted: Primates, Predators, and Human Evolution*. NY: Westview Press.

Henry, Amanda G., et al. 2012. "The Diet of *Australopithecus sediba.*" *Nature* 487: 90–93. doi:10.1038/nature11185.

Henry, Amanda G., et al. 2018. "Influences on Plant Nutritional Variation and Their Potential Effects on Hominin Diet Selection." *Review of Paleobotany and Palynology* 261: 18–30. https://doi.org/10.1016/j.revpalbo.2018.11.001.

Hlubik, Sarah, et al. 2019. "Hominin Fire Use in the Okote Member at Koobi For a, Kenya: New Evidence for the Old Debate." *Journal of Human Evolution* 133: 214–229. https://doi.org/10.1016/j.jhevol.2019.01.010.

Hosfield, Robert. 2020. *The Earliest Europeans: A Year in the Life.* Oxford: Oxbow Books.

Hublin, Jean-Jacques. 2017. "The Last Neanderthal." *PNAS* 114 (40): 10520–10522. doi:10.1073/pnas.1714533114.

Jaffe, Karin Enstam. 2019. "Primate Behavior and Ecology." *Explorations: An Open Invitation to Biological Anthropology.* Beth Shook, et al., eds. Arlington, VA: American Anthropolgica Association. 190–232.

James, W.P.T., et al. 2019. "Nutrition and its Role in Human Evolution." *Journal of Internal Medicine* 285: 533–549. doi:10.1111/joim.12878.

Ji, Qiang, et al. 2021. "Late Middle Pleistocene Harbin Cranium Represents a new *Homo* Species." *The Innovation* 100132. https://www.cell.com/the-innovation/fulltext/S2666-6758(21)0057-6.

Joannes-Boyau, Renaud, et al. 2019. "Elemental Signatures in *Australopithecus africanus* Teeth Reveal Seasonal Dietary Stress." *Nature* 572 (7767): 112–115. doi:10.1038/s41586-019-1370-5.

Jolly, Clifford. 1985. "The Seed-eaters: A New Model of Hominid Differentiation Based on a Baboon Analogy." *Primate Evolution and Human Origins.* Russell L. Ciochon and John G. Fleagle, eds. Menlo Park, CA: Benjamin/Cummings Publishing. 323–332.

Klein, Richard G. 2009. *The Human Career: Human Biological and Cultural Origins.* Third edition. Chicago: U Chicago P.

Kooij, I.A., et al. 2016. "The Immunology of the Vermiform Appendix: A Review of the Literature." *Clinical and Experimental Immunology* 186: 1–9. https://doi.org/10.1111/cei.12821.

Marcé-Nogué, Jordi, et al. 2020. "Broad-scale Morpho-functional Traits of the Mandible Suggest No Hard Food Adaptation in the Hominin Lineage." *Scientific Reports* 10: 6793. https://doi.org/10.1038/s41598-020-63739-5.

Martin, Jesse M., et al. 2020. "Drimolen Cranium DNH 155 Documents Microevolution in an Early Hominin Species." *Nature Ecology and Evolution.* https://doi.org/10.1038/s41559-020-01319-6.

Martínez, Laura M., et al. 2016. "Testing Dietary Hypotheses of East African Hominines Using Buccal Dental Microwear Data." *Plos One* 11(11): e0165447. doi:10.1371/journal.pone.0165447.

Morita, Wataru, et al. 2019. "Metameric Variation of Upper Molars in Hominoids and its Implication for the Diversification of Molar Morphogenesis." *Journal of Human Evolution* 138: 102706. https://doi.org/10.1016/j.jhevol.2019.10276.

Nava, Alessia, et al. 2020. "Early Life of Neanderthals." *PNAS* 117 (46): 28719–28726. www.pnas.org/cgi/doi/10.1073/pnas.2011765117.

Newson, Lesley and Peter Richerson. 2021. *A Story of Us: A New Look at Human Evolution.* Oxford: OUP.

Ni, Xijun, et al. 2021. "Massive Cranium From Harbin in Northeastern China Establishes a New Middle Pleistocene Human Lineage." *The Innovation* 100130. https://www.cell.com/the-innovation/fulltext/S2666(2)00055-2.

Nowell, April. 2021. *Growing Up in the Ice Age.* Oxford: Oxbow Books.

Paskey, Amanda Wolcott and AnnMarie Beasley Cisneros. 2019. "Archaic *Homo.*" *Explorations: An Open Invitation to Biological Anthropology.* Beth Shook, et al., eds. Arlington, VA: American Anthropological Association. 403–443.

Peterson, Alexandria, et al. 2018. "Microwear Textures of *Australopithecus africanus* and *Paranthropus robustus* Molars in Relation to Paleoenvironment and Diet." *Journal of Human Evolution* 119: 42–63. https://doi.org/10.1016/j.jhevol.2018.02.004.

Pinker, Steven. 2011. *The Better Angels of Our Nature: Why Violence Has Declined.* NY: Penguin.

Pobiner, Briana L. 2020. "The Zooarchaeology and Paleoecology of Early Hominin Scavenging." *Evolutionary Anthropology.* doi:10.1002/evan.21824.

Power, Robert C. 2019. "Neanderthals and Their Diet." *eLS.* doi:10.1002/9780470015902. a0028497.

Quinn, Rhonda L., and C.J. Lepre. 2021. "Contracting Eastern African C_4 Grasslands During the Extinction of *Paranthropus boisei.*" *Scientific Reports* 11: 7164. https://doi.org/10.1038/s41598-021-86642-z.

Rightmire, G. Philip. 1990. *The Evolution of Homo Erectus.* Cambridge: Cambridge UP.

Robinson, John T. 1985. "Adaptive Radiation in the Australopithecines and the Origin of Man." *Primate Evolution and Human Origins.* Russell L. Ciochon and John G. Fleagle, eds. Menlo Park, CA: Benjamin/Cummings Publishing. 257–268.

Scerri, Eleanor M. L., et al. 2018. "Did Our Species Evolve in Subdivided Populations Across Africa, and Why Does It Matter?" *Trends in Ecology & Evolution* 33 (8): 582–594. https://doi.org/10.1016/j.tree/2018.05.005.

Schwartz, Gary T., et al. 2020. "Fracture Mechanics, Enamel Thickness and the Evolution of Molar Forms in Hominins." *Biological Letters* 16: 20190671. https://dx.doi.org/10.1098/rsbl.2019.0671.

Strier, Karen B. 2017. *Primate Behavioral Ecology.* Fifth edition. NY: Routledge.

Stringer, Christopher and Clive Gamble. 1999. *In Search of the Neanderthals: Solving the Puzzle of Human Origins.* NY: Thames and Hudson.

Su, Denise F. 2013. "The Earliest Hominins: *Sahelanthropus, Orrorin,* and *Ardipithecus.*" *Nature Education Knowledge* 4 (4): 11.

Tague, Gregory F. 2016. *Evolution and Human Culture.* Leiden: Brill.

Tague, Gregory F. 2018. *Art and Adaptability: Consciousness and Cognitive Culture.* Leiden: Brill.

Teaford, Mark F. and Peter S.Ungar. 2000. "Diet and the Evolution of the Earliest Human Ancestors." *PNAS* 97 (25):13506–13511.

Towle, Ian, et al. 2019. "Dental Caries in Human Evolution: Frequency of Carious Lesions in South African Fossil Hominins." *bioRxiv.* https://doi.org/10.1101/597385.

Towle, Ian, et al. 2021. "Tooth Chipping Patterns in Paranthropus Do Not Support Regular Hard Food Mastication." *bioRxiv.* https://doi.org/101101/2021.02.12.431024.

Ungar, P.S. 2019. "Inference of Diets of Early Hominins from Primate Molar Form and Microwear." *Journal of Dental Research.* doi:10.1177/0022034518822981.

Vaesen, Krist, et al. 2019. "Inbreeding, Alee Effects and Stochasticity Might be Sufficient to Account for Neanderthal Extinction." *Plos One* 14 (11): e0225117. https://doi.org/10.1371/journal.pone.0225117.

van Casteren, Adam, et al. 2020. "Hard Plant Tissues Do Not Contribute Meaningfully to Dental Microwear: Evolutionary Implications." *Scientific Reports* 10: 582. https://doi.org/10.1038/s41598-019-57403-w.

Warren, Kerryn, et al. 2019. "Early Hominins." *Explorations: An Open Invitation to Biological Anthropology.* Beth Shook, et al., eds. Arlington, VA: American Anthropological Association. 319–373.

Welker, Barbara Helm. 2017. *The History of Our Time: Hominini*. Geneseo, NY: SUNY.

Weyrich, Laura, et al. 2017. "Neanderthal Behaviour, Diet, and Disease Inferred from Ancient DNA in Dental Calculus." *Nature* 544 (7650). doi:10.1038/nature21674.

Wood, Bernard. 2005. *Human Evolution: A Very Short Introduction*. Oxford: Oxford UP.

Wood, Bernard A. and David B. Patterson. 2020. "*Paranthropus* Through the Looking Glass." *PNAS Latest Articles*. www.pnas.org/cgi/doi/10.1073/pnas.2016445117.

Wynn, Jonathan G., et al. 2020. "Isotopic Evidence for the Timing of the Dietary Shift Toward C_4 Foods in Eastern African *Paranthropus*." *PNAS* 117 (36): 21978–21984. www.pnas.org/cgi/doi/10.1073/pnas.2006221117.

Yoshida-Levine, Bonnie. 2019. "Early Members of the Genus *Homo*." *Explorations: An Open Invitation to Biological Anthropology*. Beth Shook, et al., eds. Arlington, VA: American Anthropological Association. 374–402.

Zink, Katherine D. and Daniel E. Lieberman. 2016. "Impact of Meat and Lower Paleolithic Food Processing Techniques on Chewing in Humans." *Nature* 531: 500–503. https://doi.org/10.1038/nature16990.

5

MODERN HUMANS AND CULTURAL THEORY

This is an important chapter embodying the heart of the argument and culminates many of the book's claims. Cultural group competition and selection affect cooperation, and since selection works on variation, cooperative behaviors likely stem from clashes between cultural groups (Handley & Matthew 2020). This observation does not mean that we will always have vegans v. meat and dairy eaters and nothing more. What's proposed is the opposite, since a capacity for cooperation can conceivably arise between competing groups through hybridized seeds of cultural group selection. Innovation increases the adaptive fitness of a population. Some see an opportunity to invent; outside groups then embrace the adaptive innovation. We compete culturally, but that's where cooperative change can begin. Some social networks, like communities of vegans, can break maladaptive conformity, like meat and dairy eating, to construct a better outcome among differing groups who are willing to accept the improved change in food ecology.

Recalling the section "Education, Awareness, and Influence" in this book's Introduction, instruction of young people is the linchpin in any vision of a vegan culture. One study (Wilks, et al. 2020) reveals that children are more geared to saving many pigs over a human life than any adult is inclined. The goal of "humanity" should not be to desensitize children to other life forms. We tend to be influenced to learn from those who share our beliefs (Hilmert, et al. 2006). Of course, because of partisan politics in some countries, there are worries about who shapes the beliefs of children according to whose values. Adults, nonetheless, should not be under some obligation to inculcate children to prioritize their self-serving special humanity over all other forms of life. As we can see from many recent events, whether deforestation, habitat loss, environmental catastrophes, or pandemics, that is a very costly proposition.

To begin, graduating away from Neanderthals from the last chapter, let's now take a quick look at the ecological food culture of hunter-gatherers. Their

DOI: 10.4324/9781003289814-6

continuing way of life in many places of today's world provides, along with ape and australopith diets, points of comparison and contrast to the meat culture of most contemporary, urban societies. Bear in mind, among many industrialized nations, there's a link between high meat intake and lifestyles of low exercise and general overeating (Mann 2018).

Hunter-Gatherers and Sustainable Cycles

Hunter-gatherers traditionally ate only from natural resources, but over time this equation has changed because of external forces on otherwise remote populations. In spite of this alteration, contemporary foraging groups are used as a reference standard. Research (Crittenden & Schnorr 2016) shows how plant foods, outside of tundra-like areas and seasonal influences, account for a great part of the forager diet, and perhaps up to 65 percent than previously calculated. This conclusion is confirmed by Bonnie Yoshida-Levine (2019) who says the "bulk" of a hunter-gatherer diet is plant food. The gut biomes of hunter-gatherers, outside of modern agricultural influences, tend to be healthy (Fragiadakis, et al. 2019). According to Mark Jenike (2001) requirements of nutrition not only create behaviors but affect life quality. Unlike farmers or herders, hunter-gatherers do not exploit species or lands with unwarranted aggression and control. Modern hunter-gatherers eat domesticated products, both plants and meat, but only beginning from about 10kya. Artificially selected products are richer in carbohydrates and fats whereas their wild counterparts have higher concentrations of fiber, vitamins, and minerals. In Paleolithic hunter-gatherer models, according to Jenike, the overall majority, or 65 percent, of the diet is plant and the remainder derives from animals.

There is, consequently, better protein, fiber, and little saturated fat in the ancient hunter-gatherer diet, not necessarily including those people in arctic regions. We see here a balanced approach that has been lost in our mechanical production, preparation, delivery, and excessive consumption of meat and dairy products. Consider the linkage: hunter-gatherers understood the significance of plants well enough to settle and harvest crops from seeds (Clark, et al. 2018). From foraging for plants, nuts, and seeds, to using flora for medicinal purposes, and ultimately to farming a variety of plants and vegetables validates their vital function in the human diet. This agricultural evolution was conscious and by choice because of the human reliance on a plant-based food system.

Depending on the hunter-gatherer group the sustenance equation could vary. For example, in the North Pole the diet was almost all animal. Eating wild, as opposed to processed foods, enriched Paleolithic hunter-gatherers with more nutrients. These early people, like their living contemporaries, were also more physically active, specifically engaging in extensive walking and movement. Compare these facts to the so-called twentieth-century North American diet that is low in fiber, high in fat, high in carbohydrates, high in sodium, high in processed ingredients like sugars, and low in nutrients. Except for the cyclists in some countries, one doubts most modern humans, dependent on their autos in a car culture, get nearly the walking exercise as hunter-gatherers. Some modern African hunter-gatherers subsist on a diet of less than 50 percent meat, and this could vary seasonally, year-by-year, region-by-region. Fats are also obtained from nuts and beans, as vegans know. Among hunter-gatherers, some seasonal items are prized, like honey. As with the great apes, some hunter-gatherers, along with people in other cultures, will eat insects like caterpillars and termites. George Schaller (1964) tells the story of Batwa men who raided a honeycomb, consciously eating the grubs nestled in the wax. Other seasonal items would include roots.

Not all animals are rich in fat throughout the year since they, too, encounter seasonality with a low abundance of plant foods on occasion. Seasonal deprivation of foods can be made up, in part, by stored body fat, as with orangutans. Hunter-gatherers demonstrate that there is, as Jenike (2001) notes, a flexible "plasticity" in human digestion via adaptive omnivory and in response to changes in the profusion of food items. Even in *H. erectus*, says Bonnie Yoshida-Levine (2019), there was diet versatility similar to that seen in modern hunter-gatherers. Thus, in accord with the argument here, we can steer away from foods that are unhealthy or harmful to the environment, as with mass meat and dairy production and consumption. Like great apes, hunter-gatherers engage in sustainable living with adaptations for slow growth, slow reproduction, although impacted from disease, and reduced energy budgets. Archeology shows that ancient hunter-gatherers lived in better supplied environments. In modern times, some hunter-gatherers have as few as two offspring while others have as many as eight, but this depends on the group and environmental conditions. Not all children survive to young adulthood. Compare this with the unchecked population growth of billions of hungry modern humans in mostly industrial societies.

According to David Buss (2019), low estimates of human meat consumption far exceed meat eaten by any of the hundreds of other primate species. This confirms what was discussed in Chapter 3. Some hunter-gatherers can spend up to 50 percent of their time hunting, but this means, says Buss, that they still have a diet in excess of 35 percent plant food. Citing others, Buss says that vegans don't get cyanocobolamin or B_{12} without meat. Yet factory or processed meat is so high in fat and other artificial ingredients, including hormones and antibiotics, that it is dangerous to one's health if habitually consumed. Many modern families still build meals around meat, not vegetables. Let's not avoid the obvious: B_{12} does not occur naturally in fruits or vegetables and vegans need B_{12} enhanced food. Commercially manufactured nutritional yeast can be fortified with B_{12}, for example, and there are vegan B_{12} vitamin supplements readily available. However, as naturopath Pam Popper points out (Stone 2011), B_{12} comes from bacteria in soils, not from meat itself. Because modern plant eaters so thoroughly cleanse their foods, the nutrient enriched soil is washed away, but grazing animals raised for slaughter get B_{12}. Early hominins undoubtedly acquired, like extant great apes, the obligatory dosage of B_{12} from eating unwashed roots and tubers, and perhaps edible seaweed. We don't really need as much B_{12} as recommended, according to Popper, and adults store this vitamin in their bodies anyway.

There are theories about hunting and gathering relevant to any discussion of evolutionary psychology. "Man the hunter" sees males as the providers where meat can also act as a bargaining chip for sex, alliance, or status. As previously noted, this is still evident in living chimpanzees, mentioned again since whether one wants to admit it or not, in line with Carol Adams (2000), there is among modern humans and chimps a sexual politics to meat. There is also a gathering hypothesis that sees early stone tools, surely preceded by wooden implements not remaining in the fossil record, as provisioning roots and other plant foods. Very early stone tools, as seen with chimpanzees and monkeys, were likely used to process plant foods (Pobiner 2020). According to this theory, in contrast to "man the hunter," those digging for tubers and root vegetables would be females (Tanner 1983). At any rate, stone tool manipulation for processing vegetation likely preceded stones for butchering and hunting.

Foraging for plant foods was always an important resourcing activity. As humans moved to colder climates, hunting became more predominant. Buss (2019) says that hunting by males allowed for a division of labor not explained by the gathering hypothesis, which ignores the human gut as supposedly evolved to rely mostly on meat eating. Human gut reduction, in comparison to great apes, probably resulted when hominins began eating more easily digestible fiber foods, plant foods processed by stones, and then cooking (Crittenden & Schnorr 2016). Buss also points out that the gathering hypothesis does not explain why humans ended up in places where land plants are doubtful to grow, like the frozen north. On the other hand, one can imagine the sharing of gathered and hunted foods between males and females with an increased male parental investment following. Meat in hunter-gatherer societies is shared among non-group members (Pobiner 2020).

The man as hunter concept originated in the early twentieth century with Raymond Dart who interpreted some finds of australopiths in South African sites as blood-thirsty carnivores. Unfortunately, this inaccurate, culturally accepted perspective persisted into the later decades of the twentieth century and even in the current, popular imagination. This invalid conception of meat as the centerpiece of a human diet, especially for australopithecines, is evident in contemporary writing, apparent in an article from the popular magazine *National Geographic* that actually seems to credit Dart's judgment (Gibbons 2014) in order to endorse a meaty "paleo" diet. When many anthropologists were trained, says Ian Tattersall (2005), there were dissenters who noted that broken and scarred bones were probably moved around as the kills of predatory cats. Tattersall says field researcher C.K. Brain from the 1960s, for instance, convincingly demonstrated that puncture wounds in the skull of a young australopith matched those of a leopard who would've dragged its kill into a tree, as done to this day, or into a cave. As recounted from Chapter 4, Donna Hart and Robert Sussman (2005) argue persuasively that early humans, like all primates, were in fact not hunter meat eaters but the hunted prey of carnivores. Long before 3mya, human ecosystems were unlike those of today with considerably more mega herbivores and carnivores (Faith, et al. 2019). Ecology is a milieu of many organisms competing with each other and yet simultaneously working in mutual assistance as biological and even cultural phenomena. "Man the hunter" paints a rather simplistic picture of what's outlined in Chapter 4 covering *Australopithecus*.

An ecologist, according to Eric Pianka (1974) in his classic text, is one who tries to understand the causal effects of organisms and environment in a feedback loop. To some extent, that's what's investigated here and in recent work by Karl Niklas (2016), Matthew Fisher (2019), Clarence Lehman, et al. (2019), and Mary Ann Clark, et al. (2018). Biotas are complex because of the many interconnections and dependencies among plants and animals, from individual to population levels, interacting with soils, water tables, inorganic material, and so on. I detailed in Chapter 4 the query of environmentally limiting factors for australopiths and non-limiting factors for initial *Homo*. Early *Homo* was apparently not as environmentally specialized as kin-like australopiths, slowly migrating out of Africa to other eco-systems across the globe. Apes have fewer and larger offspring to maximize their survivability. On the extreme would be organisms in a non-competitive environment, like modern humans, where more offspring are produced and likely to survive as a hungry surplus. *Homo* emerged in a competitive environment and so developed new strategies for survival and reproductive success. In evolutionary ecology, populations live and die together based on any number of predictable or random factors and selection pressures.

Most well-fed industrialized humans, on the contrary, yearly breed and then kill billions of animals for consumption. That's not sustainable (Rippin, et al. 2021), but it is somehow culturally accepted. Pianka (1974; Fisher 2019; Lehman, et al. 2019) reports how historically the term ecological niche comprises behavior and distribution in terms of resources, predators, and competitors. So in line with cultural evolution for veganism, humans would adjust their ecological niche to exclude farmed meat and dairy. We are a cultural species in high gear, so change is more than possible. We tend to be a caring and helping species, even if that's not reflected in some corporate and political leaders. On a local level multiplied many times over, sustainable habits, like vegan agriculture, can spread. People will have to adapt to diets in new ways so as to cease the environmental degradation in any community's ecological niche.

Humans are not primary producers of food, as are plants, but are instead consummate consumers and polluters. To emphasize the point, most humans in developed countries do not use matter and energy in sustainable cycles. There's no viable communal evolution of multiple species co-evolving if industrialized humans are involved. Instead of cultivating the earth's garden of biodiversity, many modern societies have severely curbed its distribution and have shrunk it down considerably. Pianka (1974) notes how by the early 1970s it was evident that a human population explosion was placing undue strain on natural resources. There was no adequate response at that time. We've falsified the equilibrium of nature by pretending that our food source is meat and dairy coming from animal farms, slaughterhouses, or corporate agricultural laboratories. Recent authors like Matthew Fisher (2019) and Clarence Lehman, et al. (2019) impliedly ask whether human consumerist farming, at the current rate, is ecologically sustainable.

A cultural shift away from meat and dairy eating, focusing on local, sustainable farming across municipalities utilizing suburban permaculture farms, abandoned urban buildings, and Malls with green energy kitchen facilities making veggie products, could shift the environmental calculation back to our favor. Part of the social change would include, mostly, attitude adjustments about vegan culture, as outlined in the section "Education, Awareness, and Influence" of this book's Introduction. Another part would rely on refabricating existing food infrastructure to accommodate energy-efficient, small farms, like rooftop and school gardens or vertical plantations (including indoor farms), and regional distributors of minimally processed vegan food products. Gotham Greens has pioneered urban food gardening, for instance. If you own a small lamb kebab shop, you could easily prepare and promote healthy vegan foods.

Cultural Evolution, Selection, and Evolutionary Psychology

There appears to be a cultural struggle for dominance (or selection) between the corporate agriculture of meat and any vegan ecology. At this point, consumers want to think about whether those two are compatible and how their conflict plays out in cultural evolution. The dangers and rewards are both high. Tim Lewens (2015) sees Peter Richerson and Robert Boyd (2005) as posing a "kinetic" theory or "patterns of variation." Selection, for them, is not paramount. Rather, they are more to "population thinking," as is Joseph Henrich (2016). Social learning and influence frequencies, in this model, count more than any focus on the innateness of inherited Stone Age mentality seen in evolutionary psychology. These authors pointedly say that "social learning" is a strategic ingredient as humans adapted and continue to do so culturally (Boyd, et al. 2011). This mantra has been signaled repeatedly in my book, beginning in the Introduction regarding the influence of education and the arts on the awareness of young people. Population thinking and evolutionary psychology are not necessarily mutually exclusive since populations consist of anciently evolved brains and some innate archetypes and instincts. Selection can also mathematically intensify in a population any "partially-adapted forms," thus increasing variants of that trait (Lewens 2007).

Readers can see that my discussion hovers around the spread of values. Is there shared belief (via memes) or true replication and retention (via selection), asks Lewens (2015). Emphasizing cumulative population inheritance, cultural evolution may not need replicators. Alex Mesoudi (2011) might disagree, as he leans more to selection theory. In the argument here, without eliminating selection, stress is placed on human culture and not "human nature," depending how one defines human, including which ancient and possibly extant ape species. What and how we learn is embedded in our long species history, evolutionarily on both the individual and group levels. There are cognitive mechanisms enabling learning from others and the environment. Hence, there's an intersection of biological and cultural evolution.

The ideas considered concern the rise of cultural values and norms. For instance, along with the preponderance of meat and dairy, there's a definite interest in and flirtation with veganism. Adaptations not only affect current groups, but in consequence, if heritable, influence future generations. For some, the dominant food culture will remain centered on meat, while for others it should gravitate to vegan ecology. Culture is information that alters behavior, a range of these theorists cited below would suggest. What's key is that psychological inclinations and actual behaviors cannot spring from the dictates of laws but must come from the influence of education under a combination of social effects on individuals and groups. Cultural traits are not really particulate matter like genes, but are ideas linked and interacting in a system somewhat like genes, which can replicate in various ways (Lewens 2007). Even so, feelings, emotions, and thoughts occur in one's brain in neural (genetic) activity. Nevertheless, cultural evolution does not exactly form a genome, which accounts for how it can, and sometimes should, quickly shift.

If selection and culture together build adaptations in an evolutionary process that worked for *H. erectus* but, presently, not for us, we have to uncover a new model of fitness. By now, readers know that framework is in a vegan culture. To achieve our goal, theories of cultural evolution can help. I'm not offering any particular representation of cultural evolution but showing how different models can aid in promoting and spreading a healthy vegan economy.

Culture and Group Selection

As a general idea, consider how some of the papers in a classic work by Richard Lee and Irven DeVore (1968, Chapters 4 and 5) confirm that for some hunter-gatherers plants, depending on proximity to marine foods, are a large part and up to 80 percent of the diet, more so than hunted game. These findings have held up well over the years (e.g., Clark, et al. 2018; Yoshida-Levine 2019). In fact, for some hunter-gatherers, meat, when consumed consists of small mammals, birds, or insects. As noted in the Introduction here, for my purposes insects are not meaty flesh per se. Lee says vegetable food is a necessity where meat is eaten opportunistically; so why, then, *man the hunter*? The answer to that question might be more cultural than biological, it seems.

Powerful people have always wielded meat as a cultural icon of prestige, says Carol Adams (2000), marked in the European aristocracy of ages past. Peasants ate root vegetables. Meat has always been associated with masculinity and virility, evidenced in many historical texts and cookbooks, she notes. Meat is a prize and has traditionally been a symbol of prosperity (Jallinoja, et al. 2019). Of course, politics with racial overtones play into meat eating for those who have it, regularly in some industrialized countries, and those who don't, in developing countries. Meat eating was inaccurately tied to superior intelligence, says Adams, promulgated across the nineteenth and twentieth centuries, stoking colonization. She goes on to relate how meat was fed to the numerous ranks of soldiers to beef up their manly strength. For our purposes, these are all pertinent examples of how cultural influence, and not biology, can exert pressure on ideas and practices. Interestingly, thirty years after the publication of Adams' *The Sexual Politics of Meat*, much of what she says will not shock a male vegan, but that does not diminish the message. For instance, Adams unsurprisingly says that a meat economy segregates genders, divides labor, enforces women to child rearing and care, and consists of male gods and patrilineality. Plant-based cultures, on the other hand, tend to be egalitarian and feminine.

A general hypothesis is that cultural evolution is influenced by population size. According to Mark Collard, et al. (2013), the dimensions of a population have direct effects on the cultural evolution of non-industrial food production of small-scale pastoralists and agriculturalists. This conclusion is for food producers but not for food procurers like hunter-gatherers, where there is a risk difference. For the argument of this book, it's clear that with existing large populations, break-out vegan economies of local, organic veggie farms and distribution centers of nominally processed plant foods can develop in urban areas where there's a greater likelihood for the spread of environmentally friendly ideas and healthy practices. While this could happen through market forces, it's more central that education plays a role in advising interested people about the health and environmental advantages of a vegan culture. Since the emphasis is on evolutionary theory, I'm not offering a blueprint for the vegan economy per se. It's not difficult to create some forward-thinking vision and policy from the sketch offered, highlighting education, learning, and public awareness. If corporations and advertisers can generate product attention, then why can't plant-oriented nutritionists.

Some researchers demonstrate how on one island in the Baltic Sea over historical time there was a diversity of food practices independent of environmental factors with a digression from marine mammals and seafood to an emphasis on farming plants (Eriksson, et al. 2008). That model of cultural shift can be applied to solve current human and global health problems by grounding food culture in a local vegan economy multiplied many times. In most societies there are culturally evolved practices to avoid toxic foods (Henrich & Henrich 2010). Great apes, too, understand the beneficial properties of medicinal plants and mineral rich soils. These practices are transmitted through relatives, learning, and copying persons in prestige positions. Similarly, high-profile people across communities can help emphasize and culturally transmit the benefits of veganism. Learning about the paybacks of a vegan economy will disperse among individuals and groups as meat and dairy become taboo carriers of disease, parasites, pathogens, harmful chemicals, and as symbols of environmental damage and animal cruelty. Furthermore, food safety culture requires a revised, holistic approach (Manning, et al. 2019). While health hazards can occur in plant foods from dirty human handling, it's obviously more regular in products emanating from the meat and dairy industries.

Cultural transmission constitutes an "inheritance system" flowing from Darwinian evolution where institutional norms are part of transmitted traits, say Peter Richerson, et al. (2016). There's enough variation across groups, Richerson goes on, to hypothesize group selection, especially when group competitions play a role in determining advantages. That observation is crystal clear in this book. Cooperative adaptations are likely to spread from group selection, as suggested here, without eliminating individual agency. Cultural group selection is crucial within and across human societies, Richerson affirms. Cultural evolution does not necessarily supersede natural selection, kin selection, sexual selection or other evolutionary processes, but it might explain values and beliefs from group competition and variation among groups. In this way, a vegan culture is not just a proximate mechanism but can be an ultimate cause affecting evolutionary outcomes. Bluntly stated, on scale, kin selection and reciprocity are small where group selection enables widespread cooperation. Group cultural mechanisms, according to Richerson and applicable to my argument, include quick social learning, conformity, rewards and punishment, status, group symbolism, and institutionalization.

Inheritance Information and Choices

Although genes are fixed, nutrition affects physical and brain health while schooling and self-improvement affect social intelligence. We make cultural choices, asserts Kate Distin (2011). Importantly, these choices can replicate and spread in the right environment, as recommended with a vegan culture. Behavior is shaped by inherited neural and mutable environmental information where the "environment" includes the psychology of others. In many organisms, the diurnal struggle for existence is mostly one of self-interest, without counting infant care or kin relatedness. Not to place humans on a hierarchy above other creatures, but people regularly engage in self-sacrifice for others, from fire fighters to medical responders, Distin might say. Therefore, it is within human capability to step back and, looking at the big picture, understand how veganism can benefit self and others in an expanding algorithm. The question is who will act on this knowledge considering that those who grasp the panorama of a vegan culture in all its forms of personal, environmental, and animal healthfulness are physically and culturally able to make change.

For the uninitiated, veganism is a metarepresentation, or the aptitude to think abstractly about ideas. We need to get most people to align climate health with the meat on their dinner plates. At the same time, representational information has to be pondered and moved from abstraction to action, Distin (2011) might say. This switch on a large scale is feasible if aided by, for example, educators and artists who can create a functional behavioral change in whoever processes new cultural information. The more frequently reliable and positive messages of veganism are transmitted the easier it will be for people to learn and remember the representations. In Distin's terminology, the receiver must be able to process a discrete representation in order to respond. Transmission can direct prescription, as attempted in this book.

While a young adult looks a certain way because of any particular genetic makeup, she might or not inherit and alter the good/bad habits of parents or other elders. In other words, it's not just imitating results that counts. Based on her circumstances, there are matters about what she's capable of learning. For a vegan economy to succeed, people need to understand cultural intentions and achieve better outcomes. A composition of sounds, images, words, and symbols, Distin

(2011) would say, are created, shared, spread, and eventually accepted, reinforced, and disseminated widely. We can see how that composition has happened with the acceptance of meat and dairy eating, so why not for veganism. In this area, the literary, visual, and musical arts could play a pivotal role. The problem is that the vegan configuration of communication needs to be adopted not just by vegans, but adapted by tribes with other ideological values. More so, as Distin points out, new thinking is required to break old patterns; typically, intelligent nonconformists fit that mold. This is not to say a vegan culture will be limited to an intelligentsia. On the contrary, the human cognitive system, generally, can in its capacity toward conformism calibrate its beliefs under adaptive stress to accept novel conformity.

The creation of a new pattern won't happen automatically; it requires motivation and ingenuity from policy makers, leaders, and educators with foresight for a cultural revision. There are no real barriers. Even so, as suggested in Chapter 1, there will be outlier crowds who deny facts and information and why any movement toward wider veganism must be cultural. Vegan deniers can cancel information, but they cannot erase effects. Jay Van Bavel, et al. (2020), in writing about pandemic responses, show how the perception of threat, cultural contexts, views of science, and individual liberty or group cohesiveness affect social behavior. Here, too, the argument posits our current meat and dairy consumption as a health and climate plague that requires a number of solutions simultaneously. While emotions play high in this scenario based on prejudices and politics, rational behavior and arguments from persuasive community leaders and educators about moral norms are required. As Van Bavel says, zero-sum thinking, which predominates in meat and dairy agriculture (i.e., threats from vegans) is not ultimately beneficial for all.

Evolution Is Not Only By Genes

Echoing some points from Chapter 1, there are cultural tendencies in meat and dairy eaters to be verbally disparaging of ethical vegans. The explanation might have something to do with how farmers and ranchers need to protect their "livestock" from predators, invaders, and thieves. Because possession of meat was advantageous for status, alliance, finance, and sexual favor it became the social norm and outperformed, in most cultures, eating only vegetables. Think, for instance, of celebrity promotion of beef and steaks. In considering the time when people settled, there could have been both male and female cultural manipulation in a society where women valued and praised the work of animal herding by men. The meat-eating norm rapidly spread and shaped genetic variation, not only in the eating of meat and dairy, but in the widespread acceptance of that behavior, even taking into account lactose intolerance. Cultural evolution has promulgated the persistence of drinking milk from a cow intended for her offspring. Saying all of this comes by using theory from Peter Richerson and Robert Boyd (2005). Generally speaking, ours is a culturally evolved environment that predominates over veganism and favored genes to help transmit the meat and dairy culture. Darwin (1859) knew species populations carried within them information that is inherited yet variable. This is the point where genes and culture interact, combine, filter, or collide to form new combinations, we might say. At the same time, according to this theory, we can create a culture of veganism.

Across hundreds of generations, patterns form. According to Richerson and Boyd (2005), culture is how a population engages with the environment to create behavioral templates that can be shared and copied. We see how meat and dairy eating evolved in a gene/culture coevolution, and by this line of thinking a new paradigm of veganism could take hold. This is how populations adapt, which is occurring lately with a wide acceptance of vegan foods and some vegan diets. Differences of environmental care or health between groups are not genetic but socially learned, implying that they can be improved. Some individuals can exert a force of cultural change, like the American actor and committed animal rights activist Joaquin Phoenix, who advocates for authentic and not faddish veganism.

Cultural evolution is not either/or, demonstrated in how one gene increases/decreases generationally. Since there can be multiple variants in cultural evolution and transmission, including competition, we need to ponder what affects behavior in a positive outcome. Because there are vegans and many vegan products from different companies, veganism is a cultural variation striving for wider appeal and acceptance. In fact, there is variety in vegan culture. Some people label themselves as vegetarian and only avoid animal flesh; others identify as raw vegans and eat only uncooked fruit and tender greens. Ecologist Carl Safina writes on his blog that he's flexible in his eating, though he tends to vegetarianism. This is because genes don't mix and blend to nothing; they cause variety, amplified by cross-pollination with cultural variants.

At the same time, natural selection favors conformist tendencies. A current situation is the non-conformity between animal devourers and vegans. People who conform to one group can shift allegiance, and that's advocated in this book. The preference would be to see veganism rise from a subculture to a regular, mainstream, but not capitalistic, venture. Our consuming instinct has led to maladaptation regarding diet in the excessive eating of animal fats and high cholesterol dairy products. Guidelines from the American Cancer Society advocate for less red meat (Rock, et al. 2020). Just as social practices and pressures have led to these harmful excesses, cultural evolution can reduce or correct them. Richerson and Boyd (2005) remind us how legendary biologist R.A. Fisher said that male offspring of female peahens prone to the elaborate show of the peacock's tail feathers will have genes for such a colorful display as well as the genes for that (female) preference to pass on. By analogy, one might say that the vegan choice could increase the alleles that boost genes preferring plant food and the tendency to avoid animal products. This is how culture can replicate, like conspicuous consumption, a status display mechanism.

For instance, imagine all those machismo-inclined men grilling hunky steaks outdoors in the United States on July 4th. They are replicating the "man the hunter" mentality. What has changed is that we've returned to a point similar to that of our ancestors in the Pleistocene where survival decisions in a threatening environment had to be made quickly. We face global warming disaster and yet many people either shrug off the climate science or exculpate animal agriculture from responsibility (e.g., Bruno, et al. 2019; Clark, et al. 2020; Lazarus, et al. 2021). As Andreas Schmittner (2020) implies, we've altered the natural carbon cycles of land, sea, and air. Echoing Richard Dawkins' notion of the so-called selfish gene, one could add that cultural populations have their own selfish needs. If the meat-eating population is governing, those desires lord over the needs and ideals of others explicitly and obliquely. There's a clash, then, between our ancient modules of small, family groups and the larger tribes that came later. Currently, vegans are seen as the defectors from and not the helpers of the large meat and dairy eating groups. Forces of cultural evolution could shift this position while still accommodating variation. We have culture because of variation. Groups have symbolic markers that are powerful forces of inclusion or exclusion, like flags. Not everyone will become an ethical vegan, but many could simply support vegan culture for climate and health reasons, with the goal of switching diets later. Offensive and critical attitudes about vegans, once ameliorated, would be a step in wider acceptance of a vegan culture.

More restricted diets evolved through agriculture in contrast to hunter-gatherers, say Lesley Newson and Peter Richerson (2021). Our genes, in part, evolved on and were driven by that animal farming. This does not, however, preclude what's said about vegan plant foods that are also part of our culture of food ecology. Cultural similarity of small groups from our Pleistocene ancestors eroded with the advent of a sedentary livelihood grounded in animal agriculture, culminating in populous cities of hungry citizens about 8kya. In this way, we see moral aggression operating from both sides of the equation against each other. Meat and dairy eaters assert their liberties, and vegans proclaim animal rights. In order to progress and survive, these factions need to recognize the energies of biological and cultural evolution that have shaped them and point to the direction for cooperative action, a consensus.

A middle ground would be in reclaiming individual and ecological health. We can start with groups on the community level, promoting vegan awareness campaigns with gardens, and build out from there. Cooperation evolved from small bands with face-to-face interactions, seen even today among great apes. Social deception was surely shared with a common ancestor, because we see today how apes can conceal information from conspecifics (Dunbar, et al. 1999). Let's not be naïve and pretend that one day we'll all awaken as vegans. Meat is big industry. A widely used text on anatomy and physiology (Betts, et al. 2017) presents, in passing, veganism as inimical to one's health without any mention of its benefits. At the same time, meat eaters need to acknowledge the health and environmental benefits of veganism (Randolph, et al. 2020).

Darwinian Sociocultural Evolution

Akin to biological evolution, there is some type of social, graded transition in human societies, moving from gathering to hunting to herding to agriculture to industry. That's the type of impression Marion Blute (2010) offers in terms of

Darwinian sociocultural evolution. Furthermore, big industries do not have to be trusted to solve our health, environmental, and climate problems since they have been the cause. At the same time, industry will need to be involved in a vegan economy. Once on board, politicians and corporate leaders will realize that economies should not be designed to harm people financially and physically. Blute writes that sociocultural evolution, not necessarily sociobiology or gene/culture coevolution, is a process of "replication, variation, interaction" where values, beliefs, and traditions are selected to enhance group fitness. This activity produces social identities that compete with each other, resulting in descent with modification. Apply this thinking to my argument regarding the shift from harmful meat and dairy to healthful veganism. It will be interesting to see which will fail and which will prevail. There are some constraints on multidimensional selection: genes, chance, and learning. In the scene set, education plays a major role to inform young people about beliefs outside their narrow orbit. Machines and technology are not our immediate salvation. We require a revision of values that, in turn, will guide us to create technology that's harmless to animals and the environment.

With veganism, there are biological and cultural selection pressures that can enhance its spread. Existing meat and dairy production practices pose a burden, not a pathway. Meat and dairy farming are impediments to our social adaptation with their focus on ownership; animals are not property. The inertia of animal farming must be replaced by the dynamism of sustainable veggie cultivation, if that's even possible given a world population of 8 billion hungry inhabitants. Veganism as an innovation could spread successfully in biocultural ways, but it needs to compete with the continuing expansion of meat and dairy eating. From one perspective, veganism can flourish if community leaders, policy makers, and principally educators point to its health and environmental advantages over eating animals. Modern humans have constructed the factory farm model and it has, until now, structured many aspects of capital and social life. Because this model is flawed in terms of posing ill health and obesity, environmental degradation, climate change, and animal cruelty, a new cultural position of small, suburban and local vegan agriculture needs to be considered. One could argue that the relative inertia of the animal farming industry results from its function. Considering the obvious deleterious nature of this industry, its utility is becoming less functional given alternatives that can be retrofitted into most of the existing industrial food machinery. This metamorphosis is happening already with thetransfarmationproject.org helping farms convert from meat and dairy to plant-based foods. Similar transformations are finally occurring with oil and gas energy production. These developments need to be ratcheted up exponentially to a much larger scale.

Relatedly, Blute (2010) reports on research showing how suicide and obesity can be contagious in large groups. This is not surprising. If that contagion thinking works for a maladaptation, we can get it to succeed for vegan culture and re-tool social norms, first, and then meat factories, soon after. Veganism should not be an invisible subculture. Culture is, after all, learned collective *mores*, and by its nature culture is flexible. The question is how this alteration could come to pass if, as seen in the United States in 2020 during the pandemic, armed gangs in the Midwest were asserting their "liberty" not to wear masks nor abide by social distancing rules. The blood lust in open meat markets that started the pathogen's spread runs through to these groups and their assault weapons. There's an implicit analogy here. Humans acted as a catalyst to activate a deadly virus, and fringe militia groups prompted by high-profile political "leaders" infected otherwise healthy minds with a cultural disease. The aim is not to convince hostile people but to reach more even-minded areas of the population where change is likely, especially the younger generations.

Some species persist, branch off, and hybridize into varieties and new species (Darwin 1859). Others don't adapt and become extinct. This biological branching metaphor (see, also, Clark, et al. 2018) is applicable to cultural ideas and norms, too. Some mutate and diversify for better or worse, and some die off. Divergence of cultural attitudes is to be expected. If a widespread cultural practice is known to be harmful, think about why it persists. Cigarette smoking rates have gone down in some countries, but not meat and dairy eating. As Blute (2010) says, there is a certain amount of continuity in evolution since it's a process of building on what is already there, with a combination of stability and purification. This is why through visionary community leadership, policy making, and education the food systems from farming and manufacturing to supply chains and restaurants can be altered enough to make a larger cultural shift to veganism. There will be tradeoffs, certainly, but that's to be expected in any economic shift that is both sociological and anthropological. Considering the environmental failure of many capitalized industries like meat and dairy, energy, or auto, selection could favor tilting to a vegan culture. Undoubtedly, in this scenario until there's equilibrium there will be an imbalance of cooperation and conflict without collective action. Needless to say, in spite of those clinging to the imperfect status quo, a paradigm shift is necessary and justified.

Just as there is a difference between the biology of sex and the social construct of gender, so too there's a distinction in the argument for veganism. In this view, meat and dairy eating is the prevailing social construct and veganism is "biological." At the same time, a vegan culture is decidedly ascending slowly but confidently to positive reform through social thinking. Human prehistory suggests a biological trend, since, as shown, australopiths and early humans relied less on meat, like many living primates. Then, herders and farmers during the sedentary period of circa 10kya relied more on meat through new cultural norms of property acquisition, possession, financial gain and distribution, and resource control with farming. Even before the rise of cities, animal products became commodities. Nonetheless, we could make a vegan culture from our biology. In selection theory, favored genes are ones whose benefits outweigh costs. At this point in human history with massive environmental depletion because of farmed meat and dairy, to say nothing of the environmental degradation of oceans and waterways from commercial fishing, there are more advantages in switching to small, regionalized plant-based food systems. Furthermore, the more local vegetation that can be produced organically and with little energy the better, seen today in urban rooftop gardens and vertical farming. Many urban dwellers in need of work could be gainfully employed, and countless empty or abandoned lots and buildings can become a strand of pearl-like oases to complement larger suburban veggie farms.

A rebellion of sorts has to take place, but without guns. There is our genetic heritage that favors a plant diet and our more recent memetic (and mimetic) history that prefers meat and dairy. In the name of sustaining a healthy planet, one has to ponder if the gene or the meme will triumph. The meat and dairy eating meme seems more symbolic than functional at this time in most industrialized societies. The cultural theorists cited here would mostly support a gene/culture coevolution, some falling more to selection and some more to population thinking. Either way, forms of cultural evolution can move from theory to practice, in my view, and I suspect that educators and social media using storytelling and artistic images could help the cause.

Blute (2010) is correct to ask if individuals make societies or if a society determines individual behavior. Similar to a question raised in the "The Moral Complexities of Eating Meat" section of Chapter 1, choices made by scattered individuals could begin to alter the composition of society. The executives of multinational agricultural corporations should not decide which animal products you eat for breakfast, lunch, and dinner. Obviously, there's a dynamic relationship between the individual and her society, and that theme gets to the core of this writing in terms of the evolutionary case for veganism. In literary studies and philosophy we might talk of individual will against duty (e.g., a conflicted character in a novel by Jane Austen or Henry James), and we have a similar situation here concerning the agency of a person and the structure of the social group. In this book, think more along the lines of individual vegan responsibility rippling out to the group forming a community and so forth. Moral codes are interpreted differently across cultures and groups. We cannot develop a mental paradise necessary for physical and psychological health while maintaining an external society of gross consumption, bad health, and violence toward the environment and animals. These extremes do not mesh.

The Group, Hierarchy, and Egalitarianism

Natural selection solves what political economist Thomas Malthus calculated in the long eighteenth century: extreme overpopulation. We might want to contemplate existing social controls, by whom they have been developed, and on whom they are imposed. Hunter-gatherers combine among relatives, cooperate with no necessary expectation of return, and practice generosity. To counter self-interest, the group foregrounds and reminds members about moral norms. Christopher Boehm (2012) writes that communities who punish those compromising group fitness fare well. We might witness cultural revolt in line with what's submitted here, considering how much indifference to the health of life and the environment most people, keenly the younger generations, are willing to tolerate. Malthusian social and Darwinian biological selection might both run the course in fluctuating waves. Hints of this are evident already in the outcry of younger generations about climate change and the vocal push of wearing masks during the 2020 pandemic. Boehm's main tenet is that through selection a group can influence and shape the gene pool. No doubt small groups of early people were egalitarian. Hierarchical structures arose with agriculture, farming, and herding since there was property ownership with the storage and distribution of food controlled by a few. Boehm says there was undoubtedly cooperative meat sharing from the Pleistocene, with small groups, to the Holocene, with larger communities in order to minimize, but not eliminate, conflicts.

Prehistorically, in theory, groups could suppress cheaters and enable altruism. People might want to see if any of that social cooperation is playing out in developed economies where corporations and nations vie for profit and power disregarding environmental ethics. The vegan movement needs to form a counterdominant culture, it seems. As Boehm (2012) might argue, ethnocentric fears of another are ancient and probably go back to a common ancestor shared with humans and apes. This thought is related to hunting, where Boehm says egalitarian meat sharing groups replaced hierarchical alpha males. The balance then shifted in most modern industrialized societies (from democratic, socialist, and communist states) with powerful politicians, lobbyists, and businesses on the side of animal farming. As is widely known, in the American state of North Carolina alone, up to 30,000 pigs can be killed on a peak day. Those bloody guts are spewed with environmental cost. Poor air quality from producing animal-based foods causes many thousands of deaths a year in the U.S. because of livestock waste, fertilizers, emissions, etc. (Domingo, et al. 2021). Huge profits go to a handful while massive pollution and climate destruction affect many people locally and distantly.

Boehm (2012) claims moral origins lie in the adaptive design of the conscience away from bullying. Following from indirect reciprocity theory, there is little fitness advantage for the masses in corporate farming over vegan agriculture based on health and environmental losses. Many vegans are so because of a moral choice. They are ethical vegans. Big industry cannot, generally, boast of any moral good. We know from the failed examples by U.S. presidents Reagan through Bush and Trump that contrary to what is said, ballooning corporate wealth does not trickle down to the middle class, and certainly not to the poor. Rather, the money is closely held by powerful shareholders and executives as assets, investments, golden parachutes, or stock pursuant to Milton Friedman's (1970) self-satisfying doctrine of profit engorgement. This corporate income ideology is not new, evidenced in how Andrew Carnegie (1900) promoted, in his words, a "gospel of wealth." Vegans have been defanged, but not corporate animal agriculture. Social control is governed by those who devalue health (obesity) and the environment (climate change). Opposition might say that people become obese by choice. That's a fraudulent claim since corporate agriculture and fast-food restaurants have brainwashed millions of unsuspecting people to eat "food" laced with saturated fats and sugars. In many minority urban neighborhoods there is no access to healthy food. By analogy, think of the colossal deception the tobacco industry foisted onto the public or how the pharmaceutical industry cajoled millions of people, including medical doctors, with opioids.

The problem seems to be that in any functioning society the individual readily assumes the cultural behaviors he's been thrust into. In this scenario, the individual is consumed by the mentality of meat and dairy eaters. For Boehm (2012), culture is problem solving through meat sharing. Though it's now a capitalistic enterprise, on some level anyone who goes into a supermarket or fast-food restaurant for a "burger" buys the life of another creature in exchange for cash. In that insalubrious three-way economic transaction, the dead animal does not benefit from the sharing, nor does the consumer who'd be better off eating a bean burger.

As announced from the beginning, my goal is not to explore ethical theories but simply to say that even considering moral individualism, each person's intentions and actions are singularly evaluated. As a moral individual, I might never break any laws, but my lawful activity might contribute significantly to climate change. Some will argue that in a free society people should be able to do what they wish – smoke, drink, pollute, and eat meat and dairy to excess. If the structure of society includes material elements such as depths of the ocean (where our plastics end up) or in the tropical forests (cleared for palm oil plantations or cattle), then one's actions infringe not just on individual health but on the well-being of others, including wildlife and ecosystems, now and in the future.

During the 2020 pandemic in the U.S., a rather portly Midwestern lockdown protester without a mask carried a long gun and a poster that read: "My rights don't end where your fear begins." Avoiding physical harm can be a rational decision as well as an instinctual response. Likewise, trying to prevent environmental collapse is based in reason and the instinct to preserve one's habitat. In terms of selection, that protestor willfully exposed himself as well as police officers and others to a virus. He might not survive. If he prevails along with his sign and rifle, his message can become an idea others will follow as it widely infects social media. During the pandemic, many thousands of people across nations who ignored mask-wearing and other basic health rules became sick, spread the virus, and subsequently forced a huge strain on hospitals and medical care providers. By analogy, that's what meat and dairy eaters are doing to the environment.

Readers can see how this digression impacts on the underlying point about the potency of cultural evolution. Existing, scattered vegans could form into clusters making a tangible difference. The meat eating dissenters do not have to preponderate with louder voices or better media coverage. The implied question is how any intelligent, civil, global citizen will contribute to the cultural evolution of a vegan diet. Many people have learned to eat meat and dairy three times a day, but given our current global health and environmental problems, that is not a rational choice.

Customs, Consumption, and Advertising

There is a social construction of reality (meat and dairy) and the revised construction of social reality (veganism). Perceptions of reality can shift depending on gender, "race," economic status, social class, or level of education. Some of those designations are false constructions. When I identify as a vegan, people tell me how little meat they eat. I tend not to believe them. On the one hand, there is a collective symbolism in meat and dairy eating, as opposed to, on the other hand, the health and environmental facts of veganism. We need to reconstruct farming and eating away from the physical magnitudes of an obesity crisis, climate change, and animal cruelty.

What's critical is not necessarily any difference between humans and animals, but what the information consists of and how it's acquired, Blute (2010) might suggest. Consider for a moment how children are reared on meat without knowing where it comes from. Contemplate the example of a child who is fed, and enjoys, tongue or liver only to realize years later with some chagrin those are the body parts of a cow. One's perception of reality can be distorted by the social views of others, and not necessarily for the benefit of physical or mental health or the sustainability of the environment. This is to say nothing about the persecution of animals, where some authors believe animal farms should be held legally culpable for cruelty (Maerz 2020). We are raised to see cows, pigs, chickens, goats, etc. as happy "farm animals" who seemingly want to be slaughtered to feed us. On the side panel of meat trucks traveling around the N.Y.C. Metropolitan area, one sees a smiling pig as if he's looking forward to ending up in someone's stomach as a cooked meal.

The consequence of this meat and dairy ideology is that we interact with our "pets" but not with farm animals. The latter are meant to be eaten, but not the former. Meantime, there really is not much biological difference (sapience and sentience) between a cat, dog, or some conscious being bred as human food. All are self-aware creatures. We have been socially raised to distinguish between pets and animals as food. Neurochemicals and neurons in our brains are programmed to propagate that viewpoint to others, especially to children. Parents with differing dietary views are in conflict in raising children. Carlo Alvaro (2020) sees ethical veganism analogous to an uncompromising religious stance, an approach that could become part of cultural evolution. Josh Milburn (2021) views any such rigid position as extreme since it demands zero-compromise. Milburn would educate on the side of animal rights since issues of justice could fit into a child's upbringing. An ethical vegan would not eat a plate of fried chicken put in front of her any more than an orthodox Jew would eat non-kosher foods. Without taking away from Alvaro's point about adults, perhaps there could be some flexibility as children are educated to accept veganism on their own terms, suggested here. Holidays revolve around elaborate meals like a dead turkey, a dead pig, or a dead lamb. No one in most cultures bats an eye about eating "leg of lamb" garnished with mint jelly, but shudders at the thought of eating part of his pet dog. Packaged heads of dogs or baby octopus are for sale as food in some countries. In other words, let's challenge the notion that we are biologically or psychologically predisposed to eat meat, and given the evidence adduced so far, there's no good reason to force animal products onto children. Give young people the facts and information they need to make decisions about their health and ethical standards. Think about the real point behind the social construction of meat and dairy eating: agricultural and retail profits.

Evolutionary processes select genes to be filtered through generations enabling fitness or the ability to survive successfully so as to reproduce. Nevertheless, in many modern human societies, by analogy, the maladaptation of excessive meat and dairy consumption has evolved. To wit, obesity and many illnesses related to corpulence. The answer is that animal farming is a capitalist construction (even on the "family" level) where some get rich while others get sick. Citing John Kenneth Galbraith in *The Affluent Society* (1958) and *The New Industrial State* (1967), Blute (2010) asserts how after World War II corporations created product demand through advertising and marketing. Actually, there's plenty of evidence of this earlier in Victorian periodicals and the magazines and billboards that soon followed. There's also the example of the popular 1948 movie *Mr. Blandings Builds His Dream House* (based on a 1946 novel). The lead character as a representative U.S. American (i.e., bland) is an advertising executive played by Cary Grant who has his marriage, family, and career premised on how well he can sell a pork product called Wham. In the end he succeeds through his African-American housemaid with her slogan, slightly racist: "If you ain't eatin' Wham, you ain't eatin' ham!" Billions of sapient and sentient beings, therefore, are reduced to slaughter and dinner by the corporate success of Madison Avenue.

Natural selection can alter course expressly through cultural evolution, and that's what's proposed in this text by turning away from meat and dairy toward veganism. Phenotypes are plastic and can change relative fitness and population gene frequencies. Ecology plays a role here as the social environment. In the case of most modern human societies the "ecology" is what they eat, which has been badly determined by corporate agriculture. Given social variation, this damaging ecology of food can change, develop largely into vegan culture, and pass on like genetic material. Ultimately, there's a question of values. Some will favor the profitable animal business of meat and dairy that is unhealthy, harms the environment, and is cruel to animals. Others will prefer smaller, energy-efficient plant industries and local, organic produce farms. In the long-term of social progress, we know which one is more sustainable. Clearly, social education is part of this equation.

From the perspective presented in this book, some human groups and societies have increased to an unscrupulous degree by creating barriers for other species who keep the environment clean. Ocean fish sustain the waters while mammals minister to the biodiversity of tropical and rain forests. Given the climate science at our fingertips ignored by many, propaganda about progress is an illusion. Animal agriculture is as dirty and unhealthy as "clean" coal or filtered cigarettes. The issues are substantive, where we can see the data, consequences, causes, and effects of animal farming to human health, the biosphere, and to particular ecosystems and their inhabitants. Cultural information dissemination is a crucial element to future generations, just as it has been to our predecessors. How the science of healthy eating is presented to young people is equally important. Parents and educators need to think about the types of messages they want to send in that cultural material, from which evaluative issues can then evolve. Right now, it's not broadcast loudly enough that individual families are negatively affected by the ill effects of the meat and dairy businesses.

Human Culture is an Evolutionary Process

Cultural files are stored in the brain's neural connections, says Alex Mesoudi (2011). According to evolutionary psychology, those earlier hominin "memories" of small group sustainability are still coded in the archetypes of our collective unconscious. Since we are cognitively adept and intellectually flexible, the challenge is to replace the existing culture of meat and dairy eating, ingrained in the human brain for hundreds of generations, with the earlier hominin plant-based model. Worse, this neural data of meat and dairy eating is fixed not only in the brain but reinforced in cultural modes and cues, in print and visual matter, ranging from social media, books, and entertainment. This is why the graphic arts and storytelling could help educate young people about the overall benefits of a vegan culture. Institutions, too, have fixated on meat and dairy where corporate, government, and school cafeterias nod in passing to meatless options on a menu otherwise brimming in animal flesh. Culture, then, is information that affects behavior.

We are genetically predisposed to eat for survival, but we are not genetically programmed to eat meat three times a day any more than we are mechanized to eat candy bars. Of course there are some qualifications here, as Mesoudi (2011) says. East Asians lack an allele to digest alcohol. We know from the exposition of australopiths previously narrated that, if anything, our brains and bodies evolved on fruits, leaves, nuts, and seeds more than meat. Australopiths must have done something right because in a roundabout way through long survival and relatedness they gave rise to *Homo*. Culture is not automatically the explanation for our diets given our evolutionary history, and yet it is because the human diet changed away from plants to meat. Let's be careful here, since the spotlight is on gene/culture coevolution. While culture seems to prime us, we are not totally conditioned by it. Individuals and groups can make choices. During the 2020 pandemic in the United States, some armed groups promoting liberty raged against those who advocated for health preferring a longer shutdown of businesses. We learn from others and from our past actions. Broadly speaking, evolutionary psychologists lean more to genes and less to culture, but both create a result. Genes permit us to learn and make choices. Genes are responsible for capacities but not differences in behaviors, Mesoudi might say, since humans are mostly genetically similar.

Culture has the power to alter and reshape behavior with a move to veganism. Cultural evolution is not progressive and societies do not move up a ladder. Instead, there are variations within a population that, through selection mechanisms, cause change over time. Roughly speaking, specialization can occur, and it's fair to ask why we have come to rely on meat and dairy away from plant foods. There's no reason we can't effect a positive change toward extensive vegan agriculture, even with insouciant attitudes by some leaders concerning climate change and animal rights. What's the viable "adaptive landscape," Mesoudi (2011) might ask. The content bias of this book's proposal is not very attractive to most people in many cultures and perhaps why veganism is slow to spread. In some societies, notably India and some African nations, meat is expensive or hard to come by, so vegetables predominate in the food culture. Religious sanctions concerning certain foods (e.g., beef and pork) also influence attitudes toward animals in some cultures. However, the vegan camp might try to ramp up the evolutionarily inherited disgust bias and ethical positions against meat and dairy.

As noted, disgust is a biological adaptation, like fear, to keep us away from harm. At this point, it almost seems counterintuitive to rely on meat and dairy farming when that land and labor could be put to cleaner and more sustainable use. Governments subsidize cattle ranchers, so why not vegan businesses. Any such investment needs to be substantial and distributed equitably. Mesoudi (2011) says innovation is diffused in certain ways, though the fundamental foundation for switching to plant foods exists in the land, factories, and labor. These are obtainable goals partially met in the existing vegan economy. There is already a frequency bias toward veganism. The meat and dairy eating group is larger than the vegan cohorts, but if the vegan collectives can cohere better and lobby more effectively change could happen. We see small vegan clusters scattered across social media and the spectrum of the internet. Distortion of vegans is rampant since they are viewed as antisocial (MacInnis & Hodson 2017). Beginning in the nineteenth century when their movements flourished, vegetarians were depicted as cultish "freaks," and sadly that falsehood persists (Jallinoja, et al. 2019). With clear focal topics and a ground base, the health benefits of veganism could be better and more widely spread without negative bias. From one point of view, meat and dairy are, ultimately, antisocial in how their excesses promote bad health, pollution, and animal cruelty.

Culturally, people tend to behave like others without necessarily considering costs/benefits, says Mesoudi (2011). Here, the effort is in trying to show that copying vegan culture is less costly, more beneficial, and reasonable in the long run. Since it's not a predominant culture but only a subculture, it's difficult to broadcast this message. Social scientists (Finkel, et al. 2020) indicate the high degree of polarization between the two major political groups in the U.S. to the point of sectarianism and aversion of each other. Since politics has assumed cultural issues, there is "othering" of those who appear outside the perceived moral norm, like vegans. Indeed, dietary preferences seem determined by political affiliation with deep divisions (Economist 2018). Additional strands of thought emerge here. For example, Josh Milburn (2016) sees the early animal rights philosopher Tom Regan (2004) as a political and moral thinker, where the two are sometimes difficult to separate. Veganism, too, is not only an ethical position about eating animals; it could become a political stance about food, evident in the work of Sue Donaldson and Will Kymlicka (2011) on animal citizenry. Following from this line of thought, Kadri Aavik (2019) explores vegan identity and practice in Estonia as a social and cultural movement expressing deliberately disruptive independence from the dogmas of the former Soviet Union.

Culture Shapes the Human Mind

A pertinent question by Kevin Laland (2017) is whether evolutionary biology can explain some of the complex social systems, infrastructure, and products we now have. The implied answer is yes. Furthermore, as argued here, these same evolutionary predispositions, given the alteration of consumerist, capitalist, and traditionally conservative political barriers, can help us and the biosphere exist peacefully and healthfully into the next centuries. Some might say there's a gap between the accomplishments of humanity and other creatures. I would not disagree more since many organisms, whether flora or fauna, have survived longer and better than have modern humans. Rather than living in harmony with nature, we have created a chaos of destruction. It is true, nonetheless, for Laland to say that human advances developed in a coevolution feedback matrix that grew and spread exponentially. Following Griskevicius and Durante (2015), much consumer behavior is not just an individual choice; consumers are influenced subtly or overtly by others whom we know. It's common knowledge that women tend to be vegans more than men. Perhaps, not coincidentally, gun culture is also customarily male dominated. There could be evolutionary roots in this mental and physical habit, where men associate meat with masculinity, forceful acquisition (rifle hunting), and possession (a mate).

We need to tip the balance away from a spiral downward with a move to planetary vigor. We should be able to achieve the goals outlined so far through awareness campaigns and, mainly, education of young adults. If culture solved past problems, then a vegan culture friendly to health, the environment, and animals should become our next great iteration. Innovative behaviors can occur quickly, which means they are culturally learned. Since humans, like other animals, have underlying genes to enable imitation and learning, we need a widespread effort to promote veganism. At the same time, frankly, one is puzzled by people who claim human exceptionality (Henrich 2016) and a great divide between us and other species. Each species is unique per evolution by descent with modification. Other animals feel pain, grieve, and comprehend fairness, not just humans. In terms of the bottom line, among living creatures, the recent anatomically modern humans have a poor track record of environmental care. If there were a court of environmental crime, the human rap sheet would be voluminous. Laland, for instance, consistently downplays great ape intelligence, communication, theory of mind, and moralistic social behaviors as "romantic" only to enhance as superior human capacities. Laland says human achievements, in comparison with other species, are "impressive."

More is not better. Bigger is not necessarily an advantage. The human population explosion and colonization of nature's pastures and forests is not an adaptive solution but the very problem we face. Humans are not good ecosystem engineers, especially by feeding 8 billion hungry people with meat and dairy. We can try E. O. Wilson's (2016) idea of inhabiting merely half the earth leaving the remainder wild. We can turn to vegan agriculture. We can employ green energy. We can shift to small, local, organic veggie farms, whether suburban, urban, schoolyard, or vertical by repurposing uninhabited buildings, shopping centers, or decommissioned military installations. Kitchens in these structures could be managed for the production of veggie products by those seeking employment. None of these propositions have been seriously discussed by community policy makers across the board, as noted in Chapter 1. Adam Vanbergen, et al. (2020) outline some bigger ideas along these lines, while the thinking in this book is focused more on a neighborhood level where effective action is plausible. If humans are smarter than eco-friendly animals, let's see the definitive proof we have not exploited the gifts of nature's abundance. Laland (2017) says our success is in how we discriminate about what to copy. We don't simply imitate everything we see. Given the high degree of human intelligence he cites, most people keep repeating the bad behaviors of meat and dairy farming, production, and eating fueling that industry of ecological horror while ignoring veganism. Someone could advocate for sustainable development and yet dine at a steakhouse. Such dangerous contradictions don't exist in supposedly less-intelligent creatures.

The mistakes of human consumption are what has been magnified and ratcheted up into an uncontrollable mess. We see holistic behaviors, instead, in so-called lesser creatures. The current, mainstream perception is that veganism is antisocial. Natural selection seems to favor social learners, but many innovations have happened by individuals or fringe groups that are not so self-interested. Social learning gloms off what is already there and is easy. There's more of a challenge to engage with something new and perhaps costly to one's status quo persona. Under certain conditions, like the threat we now face in terms of climate change, asocial learning must increase so at some point it will tip the balance in favor of the presently unaccepted vegan diet. This plan is strategic copying, which has well served many species evolutionarily in their adaptations. Ultimately, a shift to vegan agriculture will be less costly than the polluting animal agriculture, both large and small, of meat and dairy. Copying alone is not adaptive. Rather, one must copy efficient behaviors. This means one has the ability to consider alternatives for the best outcome. Based on climate science, we know where we are headed and yet avoid practices, like veganism, that will promote global health with near immediate gains.

Selection favors innovations for survival, but many modern industrial societies resist a simple innovation like a vegan culture. Gene/culture coevolution suggests how some populations have adopted different diets. There's a feedback loop in terms of what's eaten (culture) and how it affects genes. Therefore, as Darwin (1859, 1871) knew, we can artificially impose new selection on our genes, and the best course in the right direction is for a vegan diet. Laland (2017) says gene/culture coevolution dominates human history over biological evolution. So change is attainable once cultural prejudices and biases are ameliorated. We can indeed change our minds. Culture helps us adapt, and we need to apply that principle soon. In human history culture has worked along a linear graph like so (following Laland): learning; tools and hunting; butchering; fire; cooking; expanded range; diversity for all the preceding; increased complexity of cultural practices; and so forth as the physical environment became a controlled resource and not a threat. The tide has turned against us because of our own actions. Hunter-gatherers have less cultural evolution since they are mobile, forage, and can't carry around lots of goods or equipment. Modern people in industrial societies should not view that way of life as a shortcoming. Some don't, evident in the peripheral movement toward tiny houses.

With small groups, hunter-gatherers have less opportunity, and perhaps less need, for innovation. Hunter-gatherers tend to be egalitarian with no formal social structure since all work fairly and continually, so there might be less complex cultural development. Obviously, industrial nations can't go back to this model, but some aspects of it can be copied. For instance, why not eliminate many unnecessary energy consuming products and machines like the factory farm and transportation of meat and dairy. Imagine all of the wasted, empty spaces outdoors and indoors in cities that could be used for organic farming of plants and vegetables, either for fresh consumption or crafted into cooked "meats." Reliance on a localized plant-based economy can be achieved by altering the existing community infrastructure. Agriculture has been a benefit, as Laland (2017) suggests; granted, but it enabled a population explosion while the food was controlled by a few. Like great apes, hunter-gatherers space out births with a natural control. In fact, the Anthropocene began tens of thousands of years ago when we began killing off many species by large-scale, coordinated hunting, depleting natural resources for new cities, upsetting biodiverse areas with building development and settlements, and spreading disease not only in concentrated areas but also by global travel. Through a vegan culture we can amend and correct some of these unsuccessful strategies that, certainly, were initiated by very different cultural ideologies.

Cultural evolution and human cooperation have cut several ways, not all beneficial in the long run as evidenced by the 2020 pandemic from human/animal contact and international transport. Any population's reliance on agriculture is cultural and selected for on the group level. With adjustments in attitudes our practices can change again for the better. Laland (2017) seems to suggest a link between imitation and the ratchet effect. Good imitation plays on mirror neurons, emotions, and our ability to empathize. If prominent leaders were to embrace veganism, others are likely to follow. Unfortunately, there's currently no prestige for a meat eater in the social hierarchy to copy vegan behaviors.

Humans Are Wired for Culture

We are a tribal species. Mark Pagel (2012) notes how even in communities there are many cultural units or ideas housed in neurons that can be transmitted to other brains in the tribe. We didn't choose our parental genes, but we can decide on any number of cultural practices. Over the course of human history, it was increasingly culture that offered solutions to problems and not genes alone, says Pagel. We are at a crossroads. Some of the very cultural practices that propelled our species to world conquest, like industrial agriculture in the example of feeding beef to warriors (Adams 2000), are simultaneously poised to destroy us. We are not totally free from genes if we permit some of our cultural practices to kill us. Like my genetic ancestors, I want to survive, as you do, too. Natural selection is not concerned with our current health per se. Natural selection is for survival in terms of reproductive success. Cultural evolution, on the other side, can pull us out of this health and environmental crisis. We cannot rely on natural selection since we are dealing with a catastrophe that demands an immediate solution. Switching to vegan agriculture using public awareness and education is part of the answer.

True, our immune systems can evolve rapidly, but there's no immunity against devastating climate change. As Pagel (2012) says, culture is for genes. After a poor start with disease from sedentism and close contact to animals, farming indeed helped reproductive success, but now our explosive culture of meat and dairy is not to any human advantage. Those who eat animals have seemingly developed a type of "herd immunity" to the moral complexities of this problem, an ethical and cognitive dissonance. Social learning involves deliberate design awareness in order to improve behavior. This is not chance. We have the model for improved human and environmental health in veganism. We won't be saved by machines or computers, only by changing our habits and attitudes – a cultural evolution. In terms of culture, the external is a mirror of the internal, as Pagel suggests. Slaughtering animals and massively polluting the environment reflects poorly on the consciousness (and conscience) of intemperate meat and dairy eaters. Pagel says there is a tendency among humans to split into cultural units, but that notion of separation boils down to property possession and a rigid group mentality, which can be counterproductive when working with other groups. A vegan culture can be inclusive with its sustainable ideas and "meats" that can be fashioned to accommodate diverse groups.

Estrangement from others symbolizes the desire for having gene copies nearby. Cramped tribal thinking like this can promote harmful practices for other groups. Cells on their own can die, but by forming organs in organisms they have a better chance of survival. As a biologist, Pagel (2012) offers the following metaphor. Originally there was a single thread RNA competing with similar strands. Eventually, two filaments joined for a better chance of survival and replication, thus forming DNA. Veganism is expanding, but slowly, and for many it might be a health kick or celebrity fad. This group must grow with committed ethical vegans on multiple community levels to surpass the meat and dairy eaters. Otherwise, with rival factions, there is continued poor health, climate degradation, and animal persecution. The moral of the story, perhaps exaggerated but in line with a biological analogy, is that with the crises we now face meat and dairy eaters are the antisocial cheaters who must be surrounded by the altruistic vegans. Without sounding too militant, meat and dairy eaters are responsible for environmental costs paid for by vegans.

By nature, Pagel (2012) reminds us, humans are nepotistic. This value system in many cultures has led to a focus on "humanity" at the exclusion of all other species. A self-interested approach of this magnitude has produced more harm than good. The nepotistic temperament has excluded vegans and, because of prevailing cultural evolutionary forces, continues to incline in that direction despite its noticeable failures. In a hunter-gatherer band, all must contribute. In most modern human societies, regarding a move to vegan agriculture, that's currently not the case. For meat and dairy eaters, vegans are as if another species. This is not hard to imagine as we see clashes between nations, groups, and "races" all the time where imaginary differences are manufactured simply to pit one ideology against another. In this case, there is a real difference between meat and dairy eaters and vegans, especially on the subjects of short- and long-term health and environmental consequences.

One can become partial to any cultural movement because, as Pagel (2012) says, natural selection has favored us into what Robert Trivers (2011) calls the folly of fools. We have feelings and emotions that deceive us into promoting only our group's interests. Following the prescriptions of biology, Pagel would say genes affect preferences. Based on the data provided, it's clear that humans are not predisposed to eat meat and dairy (e.g., adult lactose intolerance). Moreover, cultural evolution dictates that any assumed engraved behaviors are malleable. If, as is true, genes become more pronounced in suitable environs, more people should be gravitating to veganism. The environment of evolutionary adaptation, while still in our brains, counts for less in our culturally evolved societies, so the "environment" of cultural pressure is more paramount. As natural selection is a sieve that winnows out traits, culture sorts talents, Pagel notes. For the argument in this text, culture can be substantial movement where more are included than selected out.

The expectation is that via cultural evolution the practice of meat and dairy eating will be mostly removed and replaced by vegan agriculture. This process is likely inevitable since it will advantageously enhance survival for the biosphere. False beliefs in a group, like fears and biases, can encourage survival. Convincing friends that only meat and dairy provide protein and all essential nutrients does not qualify as a sustainable belief system. Look at a silverback gorilla who is big, powerful, and yet who is predominantly a vegetarian. Some reckless behaviors like conspicuous consumption evolved as signals of fitness. Males tend to be less risk averse than females. In spite of health hazards, both men and women conspicuously consume large amounts of red meat to signal their prosperity, individual liberty, sex appeal, or wanton disregard for temperance. Consider all the social media posts of fat-laden meaty foods. Although cooperation can be competitive, it's an investment in line with the famous kin selection equation of William Hamilton (1964), $r \times b > c$. Relatedness multiplied by benefit to genetically-like others is greater than the cost. More than thinking cooperatively, we have to act collectively, as if kin related. It's worth repeating that the vegan movement can begin with small, connected groups on the community level and from that juncture work outward.

In large part, then, let's address the question of how genes fit into an environment. If more organisms with healthy diets survive better than others, those genes (of food resourcing for survival) are passed on. At the least, and more likely for social creatures, the genes for learning from others how to forage are inherited. Mutations can change the overall numbers in a gene pool, and in a vegan culture the mutating mechanism is social and not strictly genetic. An organism is a collection of identical cells (originating in the zygote), many of which have different functions, but are all working together altruistically. Social proximity could encourage altruism, and groups are like organisms. In fact, as J. Scott Turner (2017) says, invoking Hamilton (1964) and looking at the extended evolutionary synthesis model, inclusive fitness motivates social behavior. In a vegan economy, as laid out here with community and schoolyard gardens, etc., one could see how the epigenetic material of veganism for health, sustainability, and animal ethics can be heritable through an extended social altruism for non-kin, the environment, and animals.

According to Pagel (2012), natural selection has helped us to be better at detecting deception than engaging in it. Logic dictates that people should be able to see how the agriculture of meat and dairy has deceived us into believing widely advertised slogans about the manliness of beef and the sexiness of milk. People don't necessarily know their essential character but make inferences about their temperament and personality from perceiving their behavior reflected in others or elsewhere. The hope is that many people will become more mindful of their meat and dairy consumption and the severe consequences of that lifestyle, as they observe vegans, so as to reduce or eliminate it to effect real change.

Going Vegan Without Lab Meat

Regarding cultural evolution, we need to examine the direction in which the meat industry is headed. I should note that the science of lab meat is changing rapidly, but what I recount here is still accurate.

There are a host of small companies who manufacture vegan "meats" and "cheeses." These entrepreneurs are no doubt profit-driven but are helping to improve health and save the planet in a way contrary to multinational mega corporations based in animal agriculture. Some raw vegans would eschew whole grains and cooked products. If the goal in terms of cultural evolution is to reduce our dependency on meat and dairy, then any wholesome alternatives should be considered. A dramatic shift to simplified raw veganism will not be embraced by many who are used to cooked foods and who could find edible substitutes in soy (assuming moderation and no allergy) or other fleshless products. I say all of this to demonstrate how some animal industries could re-tool their works to plant food manufacture as part of the modified cultural trend. In this way, there could be mitigation of many losses were the meat and dairy industries to shut down completely and suddenly. This change in direction is where governmental leaders and policy makers need to act, with plans and funding for such a modification. The infrastructure and knowledge exist in the many vegan companies already operating, but there's no governmental leadership to promote any forward-thinking vegan movement as espoused in this book. The legislation examined in Chapter 1 proves that limitation.

Readers can see that the argument is not wholly abstract or just some ethical "theory." It's practical with roots in human evolution. We've strayed very far from healthful practices. Critics might say an answer to the problem over which we're hovering is in so-called in-vitro lab meat, where cells are removed from a living animal and replicated in a laboratory. This lab meat could reduce animal pain and suffering and perhaps benefit the environment. Many ethical vegans are not so quick to accept those arguments. Some lab-grown flesh is still animal meat, and breeding animals to harvest their cells will increase in an intrusive manner as demand surges. This means animals will be fed grains that could have been used otherwise, will be penned, will consume resources like water, and those resources will be eliminated as polluting waste. Despite the advertising, this is not "clean meat." Lab meat makers are mostly in competition with, if not funded by, traditional meat producers. Worse, the corporate agricultural industry will seize on lab meat as a viable alternative thereby exponentially increasing the number of animals to be harvested multiplied yet again by any number of animals for other products, like their fur, bones, fats and fluids, or skin.

Assuming there is a market for lab meats, and if that escalates, the fad vegans might cease eating plant foods and possibly move back to real animal flesh and dairy. That's not farfetched since many people who are addicted to alcohol, tobacco, or drugs can become "clean" and then slip back into bad habits, so the precedent is there. This in-vitro lab meat is not without "limitations," according to Robyn Warner (2019), such as stem cell mortality, dubious nutrition, uncertain edibility, and cost effectiveness. There are also challenges regarding animal welfare, says Warner, since the lifespan of stem cells is limited and needs to be replenished from living animals by regular biopsy. Shockingly, Warner also indicates that lab or cell meat, because of projected meat demands worldwide based on population estimates, will only supplement and not replace the eating of traditional animal flesh. That's what the academic researchers conclude, but apparently that's not the public perception where some vegans reluctantly admit that lab meat might ultimately eliminate animal suffering. Warner additionally notes that cell-based meat production is not energy renewable and will waste water, with a potentially greater negative environmental impact than traditional meat production, which it won't completely replace.

The technology of cellular meat production will not save us from intemperate eating habits causing environmental damage and animal cruelty. As indicated in this book, what's needed and which can occur rapidly through cultural evolution, grounded in our environmentally friendly past and relation to apes, are changes in attitudes about the ecology of food. The focus should primarily center on altering meat and dairy culture in industrialized societies. Expensive machines will not necessarily provide the relief we need. Rather, everyday decisions multiplied among individuals along social strata about gravitating to a plant-based diet, principally from minimally processed foods, is the best answer. Ideally, these foods would emanate from local farms and workshops as repeatedly outlined, and would help employ and feed people across communities. Paul Shapiro (2018) sees a revolution forming for meat eaters in lab cultured meat, and although he advocates subtly for health, the environment, and animals, an ethical vegan might find his book ironic. Shapiro talks about a clean-meat revolution, but the rebellion started decades ago in plant-based whole foods as people boycotted meat consumption.

Earlier, I mentioned how some companies, without harming animals, are bioengineering DNA cell codes into products like yeast to mimic "animal" protein. That's a different story than mass producing real animal flesh in a lab. The Good Food Institute (Waschulin & Specht 2018) claims they can achieve cellular agriculture where encoded genetic material for an animal protein is latched onto a host (e.g., yeast or bacteria) and cultivated. This process on its face seems different than actually replicating real animal cells from a host in a lab for human consumption. Questions, both practical and moral, abound, however, about so-called "clean meat" assertions by small companies that are working with and financed by multinational animal agricultural corporations. From an ethical vegan's perspective this substance is for people who want the taste of an animal in their mouth rather than the taste of an "egg" cooked from mung beans. There's no evidence that in-vitro lab meat will be any healthier, and there's no guarantee that omnivores who relish their meat will even eat this product. Or, they will eat lab meat and farmed meat and dairy, too. Ethical vegans won't eat lab meat. Perhaps non-invasive versions of lab meat or cellular agriculture, in moderation, could help feed those pets who are obligate carnivores, like house cats. Having or not having pets is another argument.

Molecular biologist David Steele (2015) is highly suspicious of the positively optimistic claims made by in-vitro lab meat companies. For one, he says to produce this meat in a lab requires fetal bovine serum from the heart of an in utero calf cut from the mother cow who, of course, has been killed. This deed, using such "technology," must be multiplied hundreds of times to obtain the necessary amount of serum for one lab meat burger. Second, this serum is an essential part of the process due to its hormones and protein needed for cell growth, components difficult and expensive to replicate in animal-free serum. Contrary to some claims, bovine serum in this method can't be eliminated from the production that will also require much energy, water, and antibiotics. Because this particular procedure of making lab meat is so inefficient, Steele goes on to say 3D printing might be used or that since so little "meat" is rendered it could be added to existing vegan burger products. So far, photographs of lab meat reveal a paltry offering for those humans who are beef hungry.

Cell types from farm animals can be used to artificially engineer or cultivate meat (blood vessels, muscle fiber, and fats) in a laboratory, say Tom Ben-Arye and Shulamit Levenberg (2019). Though lessened, antibiotics will still be used in lab-meat production. Currently, antimicrobial resistance to antibiotics is a global concern. These authors admit that much activity around in-vitro lab meat is still in the research and development stage, and many of the verb tenses they use are conditional (e.g., "may be"). Stem cells from an animal biopsy are crucial for lab meat research. Pluripotent stem cells, simpler to produce, require gene editing, are low yielding, and likely won't copy primary stem cells, they confess. All of these deficits could impact negatively on regulatory and consumer approvals. They go on to discuss other types of cells used for lab meats, and these derive from biopsies, carcasses, or fetuses. In the scaffolding of lab-meat, plant-based proteins could be used. While animal-free biomaterials should be used, they refer to expanding use of farm animal cells. The research of Ben-Arye and Levenberg is funded by Aleph Farms, Israel, which cultivates steaks directly from animal cells.

Although some might argue that in the big picture these in-vitro meat companies have good intentions, one essay (Chriki & Hocquette 2020) concludes that lab meats are more myth than reality, even for the near future. There have been few significant advances; research and development is expensive and necessary; the diversity of meat types will be near impossible to duplicate; dysregulation, as seen in using cancer cells, is probable; and there's uncertain nutritional value. Animals will still be needed to harvest cells. Worse, consumers might view cultured lab meats as unnatural compared to a veggie burger, and some religions might shun in-vitro lab meat as non-Kosher, non-Halal, etc., they say.

Some companies like Meatable use pluripotent stem cells, likely from blastocyst embryos, as an alternative to fetal bovine serum. In fact, public sentiment via cultural evolution seems to be steering lab meat companies away from using fetal bovine serum to produce muscle and fat cells. This trend means that in decades to come, there will be less reliance on animal agriculture, a positive development. My point is that we can achieve progressive effects now by embracing a vegan culture. Reducing real meat production in forty years might be too little too late, so why wait while spending so much money and effort on research and development for a product that's not really necessary.

For decades the tobacco industry lied about the harmful effects of cigarette smoking and made huge profits. For years oil executives hid the results of their own scientific studies warning about global overheating and rising tides and reaped financial benefits. Such dishonesties have been replicated by pharmaceutical companies. These days we still hear corporate executives, lobbyists, and politicians extolling the benefits of "clean coal," as if burned fossil fuels are good to breathe. Not to be cynical, but since in-vitro lab meat has become corporate, there likely will be some deception of consumers and the general public about health risks and potential animal suffering while profits are raked in by lobbyists and legislators for executive and shareholder earnings. Consider the amount of time, money, and human power that is exhausted on perfecting lab meat when there are innumerable vegan products on the market already. The shift in cultural evolution should be to turn away from animal meat and dairy, not toward it. Community leaders and policy makers could focus on promoting infrastructure to develop more urban green farms and distribution centers for minimally processed non-meat and non-dairy whole foods and vegan options at a reasonable cost. Let's educate young people with schoolyard gardens and vegan kitchens. Bailouts are granted to corporations that degrade the environment, or their proxies, so to be fair there should be funds in a move toward reforestation, rewilding, and veganism. We cannot assume we will have a healthy citizenry and environment in the future with lab meat production.

Like Warner (2019), other authors (Bonny, et al. 2015) confess that meat industry resources cannot ultimately meet future demands and there could be much competition for traditional meat products in the form of in-vitro artificial meat. There already are many meat replacements, delicious alternatives, and healthy substitutes in plant foods emerging from companies large and small. That's a real threat to the industrial meat and dairy complex. To compete, the meat industry will need to employ a number of options, according to Sarah Bonny, et al. (2015). In addition to the in-vitro cultured lab meat, derived from lab-grown animal tissues and cells, there could be genetically modified organisms, like "enviropigs," or cloned animals. It's hard to imagine that people will eat that stuff. The answer depends on how it is marketed, and, as has been pointed out throughout the book, corporate agricultural has so far convinced most people what to eat. As for some versions of the in-vitro lab meat, Bonny says it requires a sterile environment, which is not only costly in terms of technology, but will be difficult to maintain on the large scale it is projected to require.

It seems we're moving into a culture war of food ecology, where meat and dairy traditionalists abjuring scientific data are pitted against those who want to halt climate change, foster better physical and mental health, and act as ethically as possible toward other living creatures. We require not only policy changes but social movement activism.

Placing alternative meat burgers in any form on the menus of fast-food restaurants is a mistake for the public good. It engorges corporate coffers by alluring people into places where, certainly, they will eat high fat and high cholesterol animal meat, bringing us back to obesity and all the diseases springing from corpulence. So this argument ends where it began, in a plea for healthfulness. One plant-based or in-vitro lab item on a menu otherwise loaded with fatty red meats will not alleviate the pollution caused by animal farms. Imagine, instead, in the near future fast-food vegan outlets with no lab meats. While emphasis in this book has not been entirely on advocating animal rights, one can't help but comment how the vast majority of meaty restaurants are not taking any position against animal persecution since it cuts into their profits. The attitudes against animal well-being seem insurmountable. The cognitive and ethical dissonance is so great that one will ignore baby calves separated from their mothers for slaughter while not acknowledging any parallel to immigrant children separated from their parents. Indeed, that's a controversial analogy. This is all to say that lab-grown meat might not be a viable alternative in terms of health, cost, the environment, or animal ethics given the existing culture of ethical veganism.

Let's move to a conclusion where key ideas and claims in this argument for promoting the agricultural economy of a vegan culture beginning with community education and awareness can be neatly recapitulated. You are a biological entity in an ecosystem in a biome encompassed in a biosphere. The production of your food, in contrast to the natural grazing of great apes and early humans, is mostly having a poisonous impact on personal health and the climate, to say nothing of the harms to farmed animals and wild habitats. Much of that supporting literature was cited in the Introduction and Chapter 1. We can change for the better quickly with sustainable thinking that, in turn, promotes ecologically friendly technology.

References

Aavik, Kadri. 2019. "The Rise of Veganism in Post-Socialist Europe: Making Sense of Emergent Vegan Practices and Identities in Estonia." *Through a Vegan Studies Lens: Textual Ethics and Lived Activism*. Laura Wright, ed. Reno, NV: U Nevada P. 146–164.

Adams, Carol J. 2000. *The Sexual Politics of Meat: A Feminist-vegetarian Critical Theory*. Tenth Anniversary Edition. NY: Continuum.

Adams, Carol J. 2010. "Why Feminist-vegan Now?" *Feminism and Psychology* 20 (3): 302–317. doi:10.1177/0959353510368038.

Alvaro, Carlo. 2020. "Vegan Parents and Children: Zero Parental Compromise." *Ethics and Education*. https://doi.org/10.1080/17449642.2020.1822610

Ben-Arye, Tom and Shulamit Levenberg. 2019. "Tissue Engineering for Clean Meat Production." *Frontiers in Sustainable Food Systems* 3 (46): 1–19. doi:10.3389/fsufs.2019.00046.

Betts, J. Gordon, et al. 2017. *Anatomy and Physiology*. Houston, TX: OpenStax/Rice University.

Blute, Marion. 2010. *Darwinian Sociocultural Evolution*. Cambridge: Cambridge UP.

Boehm, Christopher. 2012. *Moral Origins: The Evolution of Virtue, Altruism and Shame*. NY: Basic Books.

Bonny, Sarah, et al. 2015. "What is Artificial Meat and What Does it Mean for the Future of the Meat Industry?" *Journal of Integrative Agriculture* 14. doi:10.1016/S2095-3119(14)60888-1.

Boyd, Robert, et al. 2011. "The Cultural Niche: Why Social Learning is Essential for Human Adaptation." *PNAS* 108 (S2): 10918–10925. www.pnas.org/cgi/doi/10.1073/pnas.1100290108.

Bruno, Morena, et al. 2019. "The Carbon Footprint of Danish Diets." *Climate Change* 156: 489–507. https://doi.org/10.1007/s10584-019-02508-4.

Buss, David M. 2019. *Evolutionary Psychology: The New Science of the Mind*. Sixth edition. NY: Routledge.

Carnegie, Andrew. 1900. "The Gospel of Wealth." *Darwin*. Philip Appleman, ed. Third edition. NY: Norton, 2001. 396–398.

Chriki, Sghaier and Jean-François Hocquette. 2020. "The Myth of Cultured Meat: A Review." *Frontiers in Nutrition* 7 (7): 1–9. doi:10.3389/fnut.2020.00007.

Clark, Mary Ann, et al. 2018. *Biology 2e*. Houston, TX: OpenStax/Rice University.

Clark, Michael A., et al. 2020. "Global Food System Emissions Could Preclude Achieving the 1.5° and 2°C Climate Change Targets." *Science* 370(6715): 705–708. doi:10.1126/science.aba7357.

Collard, Mark, et al. 2013. "Population Size and Cultural Evolution in Nonindustrial Food-producing Societies." *Plos One* 8 (9): e72628. doi:10.1371/journal.pone.0072628.

Crittenden, Alyssa N. and Stephanie L. Schnorr. 2016. "Current Views on Hunter-gatherer Nutrition and the Evolution of the Human Diet." *American Association of Physical Anthropologists* 162: 84–109. doi:10.1002/ajpa.23148.

Darwin, Charles. 1859. *On the Origin of Species*. Joseph Carroll, ed. Ontario, CN: Broadview P. 2003.

Darwin, Charles. 1871. *The Descent of Man*. London: Penguin Books, 2004.

Distin, Kate. 2011. *Cultural Evolution*. Cambridge: Cambridge UP.

Domingo, Nina G.G., et al. 2021. "Air Quality-related Health Damages of Food." *PNAS* 118 (20): e2013637118. doi:10.1073/pnas.2013637118.

Donaldson, Sue and Will Kymlicka. 2011. *Zoopolis: A Political Theory of Animal Rights.* Oxford: OUP.

Dunbar, Robin, Chris Knight, and Camilla Power, eds. 1999. *The Evolution of Culture.* New Brunswick, NJ: Rutgers UP.

Economist. 2018. *"American Dietary Preferences Are Split Across Party Lines."* 22 November.

Eriksson, Gunilla, et al. 2008. "Same Island, Different Diet: Cultural Evolution of Food Practice on Öland, Sweden, from the Mesolithic to the Roman Period." *Journal of Anthropological Archaeology* 27: 520–543. doi:10.1016/j.jaa.2008.08.004.

Faith, J. Tyler, et al. 2019. "Early Hominins Evolved Within Non-analog Ecosystems." *PNAS* 116 (43): 21478–21483. www.pnas.org/cgi/doi/10.1073/pnas.1909284116.

Finkel, Eli J., et al. 2020. "Political Sectarianism in America." *Science* 370 (6516): 533–536.

Fisher, Matthew. 2019. *Environmental Biology.* Open Oregon Educational Resources/ OpenStax.

Fragiadakis, Gabriela K., et al. 2019. "Links Between Environment, Diet, and the Hunter-gatherer Microbiome." *Gut Microbes* 10 (2): 216–227. https://doi.org/10.1080/19490976.2018.1494103.

Friedman, Milton. 1970. "The Social Responsibility of Business is to Increase Its Profits." *The New York Times,* 13 September.

Gibbons, Anne. September 2014. "The Evolution of Diet." *National Geographic.*

Griskevicius, Vladas and Kristina M. Durante. 2015. "Evolution and Consumer Behavior." The *Cambridge Handbook of Consume Psychology.* Michael I. Norton, Derek D. Rucker, and Cait Lamberton, eds. Cambridge: Cambridge UP. 122–151.

Hamilton, W.D. 1964. "The Genetic Evolution of Social Behavior I and II." *Journal of Theoretical Biology* 7: 1–52.

Handley, Carla and Sarah Matthew. 2020. "Human Large-scale Cooperation as a Product of Competition Between Cultural Groups." *Nature Communications* 11:702. https://doi.org/10.1038/s41467-020-14416-8.

Hart, Donna and Robert W. Sussman. 2005. *Man the Hunted: Primates, Predators, and Human Evolution.* NY: Westview Press.

Henrich, Joseph. 2016. *The Secret of Our Success: How Culture is Driving Human Evolution, Domesticating Our Species, and Making Us Smarter.* Princeton: Princeton UP.

Henrich, Joseph and Natalie Henrich. 2010. *Proceedings of the Royal Society B* 277: 3715–3724. doi:10.1098/rspb.2010.1191.

Hilmert, C.J., et al. 2006. "Positive and Negative Opinion Modeling: The Influence of Another's Similarity and Dissimilarity." *Journal of Personality and Social Psychology* 90 (3): 440–452.

Jallinoja, Piia, et al. 2019. "Veganism and Plant-based Eating: Analysis of Interplay Between Discursive Strategies and Lifestyle Political Consumerism." *The Oxford Handbook of Political Consumerism,* Magnus Boström, et al., eds. Oxford: OUP. 157–180.

Jenike, Mark R. 2001. "Nutritional Ecology: Diet, Physical Activity and Body Size." *Hunter-gatherers: An Interdisciplinary Perspective.* Catherine Panter-Brick, Robert H. Layton, and Peter Rowley-Conwy, eds. Cambridge: Cambridge UP. 205–238.

Laland, Kevin N. 2017. *Darwin's Unfinished Symphony: How Culture Made the Human Mind.* Princeton: Princeton UP.

Lazarus, Oliver, et al. 2021. "The Climate Responsibilities of Industrial Meat and Dairy Producers." *Climate Change* 165: 30. https://doi.org/10.1007/s10584-021-03047-7.

Lee, Richard B. and Irven DeVore. 1968. *Man the Hunter.* Chicago: Aldine Publishing.

Lehman, Clarence, et al. 2019. *Quantitative Ecology.* Minneapolis, MN: University of Minnesota Libraries Publishing.

Lewens, Tim. 2007. *Darwin.* London: Routledge.

Lewens, Tim. 2015. *Cultural Evolution: Conceptual Challenges.* Oxford: Oxford UP.

MacInnis, Cara C. and Gordon Hodson. 2017. "It Ain't Easy Eating Greens: Evidence of Bias Toward Vegetarians and Vegans From Both Source and Target." *Group Processes and Intergroup Relations* 20 (6): 721–744. https://doi.org/10.1177/1368430215618253.

Maerz, Mary. 2020. "Corporate Cruelty: Holding Factory Farms Accountable for Animal Cruelty Crimes to Encourage Systemic Reform." *Animal and Natural Resource Law Review* 16: 137–170.

Mann, Neil J. 2018. "A Brief History of Meat in the Human Diet and Current Health Implications." *Meat Science* 144: 169–179. https://doi.org/10.1016/j.meatsci.2018.06.008.

Manning, Louise, et al. 2019. "The Evolution and Cultural Framing of Food Safety Management Systems – Where From and Where Next?" *Comprehensive Reviews in Food Science and Food Safety* 18. doi:10.1111/1541-4337.12484.

Mesoudi, Alex. 2011. *Cultural Evolution: How Darwinian Theory Can Explain Human Culture and Synthesize the Social Sciences.* Chicago: U Chicago P.

Milburn, Josh. 2016. "Animal Rights and Food: Beyond Regan, Beyond Veganism." *The Routledge Handbook of Food Ethics.* Mary C. Rawlinson and Caleb Ward, eds. London: Routledge. 284–293.

Milburn, Josh. 2021. "Zero-compromise Veganism." *Ethics and Education.* doi:1.1080/17449642.2021.1927320.

Newson, Lesley and Peter Richerson. 2021. *A Story of Us: A New Look at Human Evolution.* Oxford: OUP.

Niklas, Karl J. 2016. *Plant Evolution: An Introduction to the History of Life.* Chicago: U Chicago P.

Pagel, Mark D. 2012. *Wired for Culture: Origins of the Human Social Mind.* NY: W.W. Norton.

Pianka, Eric R. 1974. *Evolutionary Ecology.* NY: Harper and Row.

Pobiner, Briana L. 2020. "The Zooarchaeology and Paleoecology of Early Hominin Scavenging." *Evolutionary Anthropology.* doi:10.1002/evan.21824.

Randolph, Delia Grace, et al. 2020. *Preventing the Next Pandemic: Zoonotic Diseases and How to Break the Chain of Transmission.* Nairobi, Kenya: United Nations Environment Programme.

Regan, Tom. 2004. *The Case for Animal Rights.* Updated edition. Berkeley, CA: U California P.

Richerson, Peter J. and Robert Boyd. 2005. *Not By Genes Alone: How Culture Transformed Human Evolution.* Chicago: U Chicago P.

Richerson, Peter, et al. 2016. "Cultural Group Selection Plays an Essential Role in Explaining Human Cooperation." *Behavior and Brain Science* 39, e30. https://doi.org/10.1017/S0140525X1400106X.

Rippin, Holly L., et al. 2021. "Variations in Greenhouse Gas Emissions of Individual Diets." *Plos One* 16(11): e0259418. https://doi.org/10.1371/journal.pone.0259418.

Rock, Cheryl L., et al. 2020. "American Cancer Society Guidelines for Diet and Physical Activity for Cancer Prevention." *CA: A Cancer Journal for Clinicians* 70: 245–271. https://doi.org/10.3322/caac.21591.

Schaller, George B. 1964. *The Year of the Gorilla.* Chicago: U of Chicago P, 1988.

Schmittner, Andreas. 2020. *Introduction to Climate Science.* Corvallis, OR: Oregon State University.

Shapiro, Paul. 2018. *Clean Meat: How Growing Meat Without Animals Will Revolutionize Dinner and the World.* NY: Gallery Books.

Steele, David. 2015. "*New Meat Alternatives Offer Great Promise.*" Vancouver Humane Society14 October.https://vancouverhumanesociety.bc.ca/posts/new-meat-alternatives-offer-great-promise/.

Stone, Gene, ed. 2011. *Forks Over Knives: The Plant-based Way to Health*. NY: The Experiment.

Tanner, Nancy Makepeace. 1983. "Hunters, Gatherers, and Sex Roles in Space and Time." *American Anthropologist* 85: 335–341.

Tattersall, Ian. 2005. Foreword. *Man the Hunter* by Donna Hart and Robert Wald Sussman. NY: Westview Press. ix–xiii.

Trivers, Robert. 2011. *The Folly of Fools: The Logic of Deceit and Self-Deception in Human Life*. NY: Basic Books.

Turner, J. Scott. 2017. *Purpose and Desire: What Makes Something "Alive" and Why Modern Darwinism Has Failed to Explain It*. NY: Harper One.

Van Bavel, Jay J., et al. 2020. "Using Social and Behavioural Science to Support COVID-19 Pandemic Response." *Nature Human Behaviour* 4: 460–471. https://doi.org/10.1038/s41562-020-0884-z.

Vanbergen, Adam J., et al. 2020. "Transformation of Agricultural Landscapes in the Anthropocene: Nature's Contributions to People, Agriculture and Food Security." *Advances in Ecological Research*. David A. Bohan and Adam J. Vanbergen, eds. Academic Press. 193–253.

Warner, Robyn D. 2019. "Review: Analysis of the Process and Drivers for Cellular Meat Production." *Animal* 13 (12): 3041–3058. https://doi.org/10.1017/S1751731119001897.

Waschulin, Valentin and Liz Specht. 2018. *Cellular Agriculture: An Extension of Common Production Methods for Food*. The Good Food Institute.

Wilks, Matti, et al. 2020. "Children Prioritize Humans Over Animals Less Than Adults Do." *Psychological Science*. https://doi.org/10.1177/0956797620960398.

Wilson, Edward O. 2016. *Half-Earth: Our Planet's Fight for Life*. NY: Liveright Publishing.

Yoshida-Levine, Bonnie. 2019. "Early Members of the Genus *Homo*." *Explorations: An Open Invitation to Biological Anthropology*. Beth Shook, et al., eds. Arlington, VA: American Anthropological Association. 374–402.

CONCLUSION AND SUMMARY

Crossing Over to Adopt a Vegan Culture

By now, presumably readers can see that what's imagined is a You Too movement with collective action – you too can be vegan. We need people who can culturally transform the world, not just those tinkering with incremental changes. There are no boundaries between health/illness and sustainability/climate ruin, only choices about which side of the equation one desires. There are no barriers to sustainable farming for global health. Philosopher Marjolein Oele (2020), who also holds a medical degree, presents an argument for what she calls e-co-affectivity. In Oele's vision humans need to shift attention away from themselves and their perceived needs to feel with and participate in a community of animals, plants, and even the soil. The earth is not ours to take but to share, evidenced in how Oele says our skin is the connective interface to our emotions and material objects around us in an environment of affective interdependence, not dominance.

As has been relayed, we can advance and persevere with a turn to local farming, community vegetable planting, and schoolyard gardens. Oele (2020) affirms, along with Isabel Rimanoczy's (2021) mindset of sustainability, how the biggest hurdle we must overcome is not in finding some technological magic wand to scrub carbon from the air or deflect sunlight, but in varying our emotional attitudes and values. If a majority of people accept and adjust to a vegan culture, many of our health and climate issues will improve. A vegan culture is neither all-or-nothing nor utopian. Massive changes toward veganism are happening with discussion or action on the level of academe, social and governmental agencies, and corporations. From a health and climate perspective, a vegan economy, which already has roots across social media and food producers, is just a practical decision and needs to be accelerated. When considering what's relevant in terms of health and climate stability, the choices should not have to be difficult.

DOI: 10.4324/9781003289814-7

Food and Ethics

Writing about food ethics, Paul Thompson (2016) says that food safety, environmental impact, and animal well-being are not the only concerns. There's a social component to food ethics in terms of farm subsidies, farm workers, and food distribution. Nathan Nobis (2018) also discusses the ethics of animals as food. For a cultural movement, many people are at least aware of the politics of food, whether those people are activists or consumers. A key question is how much media attention is truly given to the deleterious effects on health, the environment, and animals because of the current food industries. Rather, televised news is filled with commercials for automobiles, pharmaceuticals, and corporate agricultural products. Contrary to what political conservatives might say, but for National Public Radio and the like, mass media is not liberally open-minded or progressive about food ethics, except for an occasional nod when a marketable story arises. While meat and dairy farming pose health and environmental risks considering the magnitude of scale, there are, nonetheless, minority voices within some modern societies advocating for change.

Concerning social and cultural epigenetics, Pim Martens (2020) uses the expression "ethical ideologies" regarding views of animals based on demographics like geographic location, personality, age, and gender. For instance, some people are idealistic about animal welfare and wish for the best possible outcomes according to moral norms, while others are relativistic and stress personal beliefs over universal *mores*. These oppositions present a dilemma in establishing a vegan culture based solely on animal rights, which is part of, but not the entirety of, this book's argument. The bottom line is that if people accept how good behavior like veganism results in positive consequences, such as better personal and climate health, then those people can exhibit ethical ideas about the treatment of animals. Ethics philosopher Michael Nelson (2021) collaborates with ecologists and says most people abstractly attribute worth to nature but don't express that value in their lives.

Some might claim that the argument for a vegan culture is unethical and includes a strict baseline with no flexibility. Clearly, the views thus offered are not extreme since the argument, simply, is that people amenable to vegan health and ethics join in and support efforts to spread the practice locally and then widely. No demands are made on anyone; choice is still on the table. An expectation is that after having read this book, a decision will be made to engage in veganism on some level. While no diet is perfect, one can endorse a vegan economy that initially tolerates some occasional exceptions. The final goal is to be a consistent ethical vegan with practical outcomes in a like-minded community that will flourish.

Symbolic Behavior of Meat Eating

As noted by Felipe Fernández-Armesto (2011), the human digestive system is more "rudimentary" than that of other primates. Humans do not have strong lips or jaws, and we supposedly eat more widely than our cousin species. Chapter 3 herein, however, shows how variety craving great apes live well with little to no meat. Compared with other primates, humans are "dedicated carnivores," says Fernández-Armesto, eating from very large animals whereas chimps consume meat only rarely since it is a "paltry" part of their diet. In fact, for many human societies, carnivory is shockingly at least one-third and up to one-half of the diet. Fernández-Armesto approaches these facts from an unfazed perspective, claiming that hominin meat eating encouraged group thinking and solidarity, with socializing from hunting, meat sharing, cooking, and eating. This inference is too simplistic and perhaps even anthropocentric. Sophisticated behaviors on par with a species' habitat and adapted needs can occur without meat. Many great ape species, as well as monkeys and other animals, like elephants, dolphins, or wolves, whether meat eaters or not, have complex social structures. It's also fallacious to suggest we need to eat meat because we adapted more gracile jaw morphology and muscles than apes, for we can easily eat raw or cooked vegetables and vegan products as well. We are flexible omnivores. Human gut simplification came from a modification to easily digestible, processed, and cooked foods, but one does not have to assume that "food" consisted only of animal meat (Crittenden & Schnorr 2016).

Fernández-Armesto (2011) also asserts that for humans, meat eating added growth of dexterity and brain expansion, but any study of great apes (Herzfeld 2017; Tague 2020) demonstrates that our ape cousins are sophisticated in cognition, intelligence, social behaviors, and culture. Hand dexterity in primates has more to do with arboreal locomotion. Great apes don't consume large quantities of meat, if much at all. In fact, the semi-solitary orangutan and the nearly vegetarian gorilla are very able physically, mentally, and socially. Fernández-Armesto goes on to say, on scant evidence, that meat eating enabled the human imagination, as if we could not live on whole grains, plants, nuts, and roots alone. On the contrary, general intelligence evolved from adaptations related to changing environments and not predominantly from carnivory or sociality (Holekamp & Benson-Amram 2017). As Carl Safina (2015) seems to suggest, tools for butchering, preparing, and cooking foods need not necessarily imply intelligence or survivability. Humans cannot live without tools of any sort, whereas most animals can, so our tool dependency could be seen, by a gorilla, as a disadvantage. Our near total reliance on tools, especially for most people who are not engineers designing them, says little about human "intelligence," just cultural acceptance of automobiles, flat screen televisions, and smartphones.

Surely the brain power of some individuals went into creating tools, but millions of unimaginative people use that technology for silly purposes, like posting pictures of their meaty dinners on social media. On the other hand, if placed in the wilderness, most technology- and comfort-dependent humans would die. There are serious questions about how technology would save us, if at all. We don't really need in-vitro lab meat, considering the vast expense going into the research and development of that "food," since sustainable vegan products are currently available. If we have a solar-powered, electric bulldozer, we'll still trample down forests.

We don't need better machines to feed and then more quickly kill larger numbers of animals for the continuing demand of fleshy foods. Consuming vast quantities of meat and dairy by virtue of technology is not the way forward. A key component to the solution is in virtues like temperance, self-control, and justice for animals and the environment. As argued, we will save and reclaim environmental health by restoring the vegan culture similar to our australopith ancestral relatives, a simpler and healthier diet. Developed societies in Asia, Europe, and all the Americas need a change in values, beginning at the community level and through honest and fair education of young people about the ills built into animal agriculture.

Very few individual humans who use clothing, cars, computers, or processed foods know how to produce any of that. Of course, cooking as a human-only function energizes food over a raw diet. Fire and cooking, however, are recent developments, perhaps in early manifestations from 800kya in the Middle East. Robin Dunbar (2014) says cooking meat, to extract up to 50 percent of nutrients while reducing feeding time, might even date back to *H. ergaster* at about 1.8mya and is still part of many hunter-gatherer diets. Raw meat if over consumed emits protein poisoning. Curiously, most authors who emphasize cooking assume it was only meat and not vegetables on the fire. As evident from the preceding chapters, there were many primate and hominin species among our ancient relatives and ancestors who survived without cooking and without primarily eating meat, as is still true today. To say that cooking helped evolve *H. erectus* disregards the linkage of vegetarian species in the matrix that preceded, lived among, and anticipated anatomically modern humans. Granted, cooked plant food, not just roasted meat, in the human digestive system requires less energy to process than raw food in an ape's. Some human societies, in fact, glorify gluttony. For example, at New York City's Coney Island boardwalk during the summer there's a hot dog eating contest to see who can consume the most beef frankfurters in the shortest amount of time.

Gross over consumption of resources does not make humans superior to other primates. Part of the problem for humans, it seems, is that we ascribe meanings, symbols, and rituals to meat eating not prominently seen in other species. Certainly, our great ape cousins share meat and trade it for favors, but not to the degree in, for example, industrialized Western cultures. For Thanksgiving Day millions of turkeys are slaughtered and eaten. For Easter millions of lambs are slaughtered and eaten. For the U.S. day of independence celebration in July millions of pigs and cows are slaughtered to be barbequed as franks and burgers. Other cultures similarly have meat-centric holidays. Corporate lawyers and executives in major cities from the U.S. to Japan customarily wine and dine clients at fancy steak or sushi restaurants. What humans call social food is really the persecution and slaughter of sentient and sapient animals, including those from fresh or salt water, with substantial environmental degradation and health risks.

There's something political in gruesomely collective behaviors grounded in a sense of privilege; a distorted ethics. Social gatherings around dead animals in order to consume them, as opposed to a human funeral, are merely signs of conspicuous consumption, a show of prestige pretending to be a custom. Young vegan artists should be encouraged to offer a counter culture narrative. As was known from early times, circa 10kya in the Eastern Mediterranean where sedentary lifestyles over hunting and gathering began, farmed animals carry disease, prevalent to this day and why "factory animals" are pumped up with antibiotics. Hunter-gatherers were already familiar with seed dispersal, plant storage, and animal husbandry during times of stress on water and supplies, but they did not create a rapacious, aggrandizing, competitive economy around such practices. From Paleolithic times to the present there is a correlation between the amount and quality of consumed food and class status. The underprivileged are always poorly nourished whereas the upward climbing bourgeoisie and wealthy sometimes eat exotic, endangered animals that spread viruses causing pandemics.

Short- and Long-Term Goals

Let's consider what we want to achieve. Evolutionary ecologist Eric Pianka (1974) writes about an ethical equilibrium, and even though it's almost five decades in coming, that's what we need in food ecology and distribution. Andreas Schmittner (2020) talks about climate symmetry scientifically and how that balance if offset will disrupt plant food sources. We began destroying nature's stability in the industrial revolution, and now we collectively need to restore harmony to nature. That's easier said than done, but individual action with the proper grassroots social media campaigning could make some difference as it grows in scale across communities. Isabel Rimanoczy (2021) offers principles of a sustainability mindset, including, for example, eco-literacy, reflection, and mindfulness. Evidence of that emotional intelligence is appearing. These are not insurmountable goals. To start, the encouraging news is how more people are embracing some form of occasional plant or reduced meat diets. Young people are concerned about the future of their planet. We need to do a better job informing and educating them about cooperating with and expanding a vegan economy.

According to Daniel Lieberman (2014), humans are mainly adapted, it seems, to produce offspring, and we are quite successful in that regard, crowding out other plants and animals. We are not, however, necessarily adapted to live long, healthy, happy lives. We evolved adaptations to live in a number of environments eating a range of different foods, and that has been the story of human evolution. We are still evolving, mostly through culture. In addition to looking at our cultural adaptations, we should focus on those that help us survive in health and happiness. Through cultural evolution we could achieve some level of mental and physical fitness. We don't need to eat intemperately the wrong foods, like meat and dairy, become ill, and then pass along those bad habits to our children. We don't need to eat lab meats. Biological evolution relies on chance variation, random mutations, competition for survival, and inheritance of fitness traits. We can duplicate that culturally. As Lieberman correctly intimates, cultural evolution depends on choice and intention, often with a goal in mind. We can ethically choose what to eat, based on how it is manufactured and distributed, with the aim of stemming poor health, malnutrition, climate change, and any form of animal cruelty.

No one becomes vegan overnight. Rather, there's a slow process of acclimating the body to raw and cooked fruits and vegetables, nuts and seeds, whole grains, along with some minimally processed vegan products out of a small, energy efficient factory. If there's an empty or abandoned building in your neighborhood, why not petition local leaders to establish a vegan food cooperative. This diet needs to be the wave of the future for human and planetary health, and with the proper efforts by many people exercising cultural evolution, it's attainable. So the short-term goal is for a consensus about the problems we face and how to solve them. Starting points could reside in community and schoolyard gardens with kitchens, etc. Then, many people can pivot without too much trouble to an easy and accessible vegan diet. Finally, those expanding community groups can apply pressure to corporate and political leaders to effect policy changes that benefit everyone, not just the rich and powerful few, into a vegan culture.

Culture affects biology by creating phenotypic variations in an extended evolutionary synthesis. Like other animals, particularly our near relations in primates (Fragaszy, et al. 2017), humans learn by traditions, which can and often do evolve. There are some actions and choices with which we are not born, like farming animals to eat them or "culturing meat" in a laboratory. Some eating behaviors are imposed on us socially. Individuals and groups, however, can constructively develop and reciprocally cause adaptive change through plasticity and niche building with cultural evolution. Many of the social and environmental niches humans have constructed over the past several hundred years, from animal farms to concrete jungles, are not ecologically viable. The concerns and struggles of Wangari Maathai (2006) and her Green Belt Movement are instructive. From the 1970s onward, Maathai worked with many women to preserve natural resources, reforest depleted areas, expand urban vegetation as well as healthy, community vegetable foods. The ecofeminist Vandana Shiva is also a fighter for the underprivileged, and her ideas about community control over resources could be copied. As demonstrated in previous chapters, we are preadapted to plant diets and sustainable ecosystem habitats and can relearn how to activate those inherited traits. Urban, organic veggie gardening and vegan food production centered on unused government buildings and lots or in schools could mimic Maathai's successful movement.

Growing Food

Just as cultural practices (grain, dairy, or rice farming) affected genes in various parts of the world, a higher resolution of veganism could increase altruism as a consequence of empathy for the health of others, toward the environment, and for animals. Joseph Henrich (2016) spends a lot of time talking about cultural adaptations (e.g., cooking, hunting, and food processing) to eating animal flesh and seafood. Another view is that we are on the cusp of a new and important cultural adaptation moving us away from the harms of animal farming. Amene Saghazadeh, et al. (2019) admit that "intuitions" for food are largely cultural, triggered by age, gender, family, learned preferences, and lifestyle. Flavors are used to enrich food consumption, but studies of animals, these authors say, show they eat what's best for their nutritional requirements, even in different locales, to correct nutrient deficiencies. This is evident in apes who eat certain plants and soil. Most modern, industrialized humans do not seem to have the same intuitions about the ecology of food as great apes.

Many of earth's people are vegetarian by necessity and why I qualify assertions with jabs at industrialized societies. There are issues of privilege here, since affluent Asians and Westerners seem, mostly, to be having discussions about veganism when it has been practiced in various forms over generations by Buddhists, Thai monks, and Jains. The vegan culture and public awareness proposed, therefore, is really geared toward wealthy European, North American, and Asian nations. Typically, the biggest meat eaters and polluters are by far the U.S. and Australia, though including a host of other countries like those in South America, Israel (OECD 2019), but increasingly China (OECD 2017). Education, especially of young people, about veganism will be vital moving toward human and environmental health. In this regard, the arts and humanities could be major players. Rather than movement in countries like the United States to defund endowments that promote the arts, such financial support should be increased, with specific attention toward projects promoting the overall health, environmental, and animal friendly benefits of a vegan culture. In a similar way, legislative leaders from the local to the federal level, in all nations, need to become vocally vegan minded, shifting away from the corporate money model and instead subsidizing organic community vegetable farming and local vegan food preparation. In-vitro lab meat should be curtailed at all costs since it's technically still meat, harms animals in most forms of production, and is geared toward financial gain for executives and shareholders.

Likewise, some of the plant-based "burgers" that closely mimic the smell, look, and texture of meat should be avoided. Those products could be a gateway food, conceivably, for meat eaters to become vegans. However, if the purpose is to make something that tastes like "real meat," then perhaps one's intention is to eat what tastes familiar, like animal flesh. Those meat look-alikes are sold in restaurants that promote meat and serve other high calorie "foods" loaded with bad fats and cholesterol anyway. Many vegan meat substitutes already exist in tofu, seitan, tempeh, lentils, chickpeas, mushrooms, legumes, bean burgers, etc. Products like those, but not limited to them, could be supported through local, energy efficient manufacturers and distributors linked to small, regional farms and cooking facilities, not mega corporate entities. The empty spaces, both indoor and outdoor, are there in the form of many city owned properties, like abandoned buildings with kitchens or vacant acreage with storage that could be put to practical use. Young minority and women entrepreneurs should be encouraged and funded as energizers of vegan fares and culture.

The bottom line is that we have to start acting locally and collectively about smarter ways of producing and processing plant foods, not farming animals as food or growing bodily muscle and fat in labs. Wildlife and their ecosystems must be protected to further stem climate change and promote healthfulness for all life forms. All the ideas recommended are attainable and worthy goals given the mental and physical plasticity of humans through collective, cooperative cultural evolution.

Summary

We evolved from plant-eating species, and our primate cousins remain committed to a mostly plant-based diet. Humans have dominated the earth because they have no special adaptations, like insects, to a particular locale. Rather, from about 10kya humans have adapted to their environs by capitalizing from the habitats of others – fishing, hunting, and then razing forests to farm animals as food. Some people would say this ecological dominance was a successful strategy to diversify populations across the globe. However, the pace at which humans have engineered the environment solely to their advantage, whether through oil and gas drilling, massive deforestation for cattle ranches or palm oil plantations, is not advantageous for the health or longevity of earth's inhabitants.

The good news is that because humans are a cultural species on a grand scale, we can alter beliefs and values to redesign food production. In fact, because there already are vegetable farms, food factories, and product distributors, the infrastructure for a vegan economy is in place. We are excellent social learners, so a shift in attitude to a vegan culture will help humans conquer poor health and a diminished environment spiraling into irreversible climate change. Humans recognize social groups and norms, fairness and punishment, so we have the psychological capacities for cultural evolution. Natural selection has shaped our cognition to observe and copy not only workable tools but also intelligent behavior that enhances survival. The human brain can adapt psychologically to a new cultural paradigm, so the saving grace of veganism is feasible for groups that, in turn, can influence others.

What you eat is a political statement. If you call yourself a progressive or liberal thinker, reconsider eating animal flesh. If you call yourself a feminist, avoid drinking cow's milk. If you call yourself a vegan, shun in-vitro lab meat in boycott. Although we have culturally accepted the maladaptive behaviors of meat and dairy farming we can, nonetheless, use cultural evolution, our primary adaptive mechanism, to direct our habits and values toward the healthful and environmentally considerate ethos of veganism. In order to understand one's behaviors and thinking, it's often useful to study others, as through this book. No one is born vegan: that's an ethical, rational choice made *after* having eaten animals. Omnivores could gain a better perspective of themselves if they were to peer at the world from the vantage point of a plant-based eater, a carnivory convert.

References

Crittenden, Alyssa N. and Stephanie L.Schnorr. 2016. "Current Views on Hunter-gatherer Nutrition and the Evolution of the Human Diet." *American Association of Physical Anthropologists* 162: 84–109. doi:10.1002/ajpa.23148.

Dunbar, Robin. 2014. *Human Evolution*. London: Pelican.

Fernández-Armesto, Felipe. 2011. "Food." *Deep History: The Architecture of Past and Present*. Andrew Shryock and Daniel Lord Smail, eds. Berkeley, CA: U California P. 131–159.

Fragaszy, Dorothy M., et al. 2017. "Synchronized Practice Helps Bearded Capuchin Monkeys Learn to Extend Attention While Learning a Tradition." *PNAS* 114 (30): 7798–7805. www.pnas.org/cgi/doi/10.1073/pnas.1621071114.

Henrich, Joseph. 2016. *The Secret of Our Success: How Culture is Driving Human Evolution, Domesticating Our Species, and Making Us Smarter*. Princeton: Princeton UP.

Herzfeld, Chris. 2017. *The Great Apes: A Short History*. Kevin Frey, trans. New Haven: Yale UP.

Holekamp, Kay E. and Sara Benson-Amram. 2017. "The Evolution of Intelligence in Mammalian Carnivores." *Interface Focus* 7: 20160108. http://dx.doi.org/10.1098/rsfs.2016.0108.

Lieberman, Daniel. 2014. *The Story of the Human Body: Evolution, Health, and Disease*. NY: Vintage.

Maathai, Wangari. 2006. *The Green Belt Movement*. New Revised Edition. NY: Lantern Books.

Martens, Pim. 2020. *Sustanimalism: A Sustainable Perspective on the Relationships Between Human and Non-human Animals*. De Biezen, Netherlands: Global Academic Press.

Nelson, Michael Paul. 2021. "How Science is Making Me a Better Philosopher." *The Philosopher's Magazine* 95(4): 31–36. https://doi.org/10.5840/tpm202119588.

Nobis, Nathan. 2018. *Animals and Ethics*. Open Philosophy Press.

OECD. 2017. *Organisation for Economic and Cooperative Development: Agricultural Outlook 2017–2026*. Paris: OECD Publishing. https://doi.org/10.1787agr_outlook-2017-en.

OECD. 2019. *Organisation for Economic and Cooperative Development: Agricultural Outlook 2019–2028*. Paris: OECD Publishing. https://doi.org/10.1787/agr_outlook-2019-en.

Oele, Marjolein. 2020. *E-Co-Affectivity: Exploring Pathos at Life's Material Interfaces*. Albany, NY: SUNY P.

Pianka, Eric R. 1974. *Evolutionary Ecology*. NY: Harper and Row.

Rimanoczy, Isabel. 2021. *The Sustainability Mindset Principles*. London: Routledge.

Safina, Carl. 2015. *Beyond Words: What Animals Think and Feel*. NY: Picador.

Saghazadeh, Amene, et al. 2019. "Intuition and Food Preferences." *Biophysics and Neurophysiology of the Sixth Sense*. Nima Rezai and Amene Saghazadeh, eds. Cham, Switzerland: Springer. 305–316.

Schmittner, Andreas. 2020. *Introduction to Climate Science*. Corvallis, OR: Oregon State University.

Tague, Gregory F. 2020. *An Ape Ethic and the Question of Personhood*. Lanham, MD: Lexington Books.

Thompson, Paul B. 2016. "The Emergence of Food Ethics." *Food Ethics* 1: 61–74. doi:10.1007/s41055-016-005-x.

INDEX

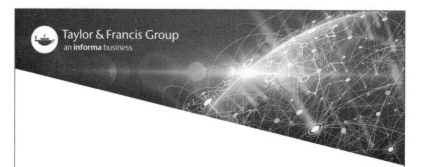